Dumpster Diving

[illegible]
[illegible]
[illegible]
Streetsmarts 3
[illegible]
Trip 3
[illegible]
[illegible]
[illegible]
[illegible]
Mail drops
Repossession
[illegible]
Criminal Law Handbook
[illegible]
[illegible]
Down But Not Out
[illegible]
[illegible]
[illegible]
[illegible]
[illegible]
[illegible]
Vital Targets
[illegible]
[illegible]

DISCLAIMER AND WARNING

The author, publisher, and distributors of this book disclaim any responsibility for the use or misuse of information contained in this book. If you choose to dumpster dive or perform *any* of the activities described in this book, you do so at your own risk. You have been warned.

The author wishes to make known that this book was written prior to the moderating influences of law school.

DUMPSTER DIVING

THE ADVANCED COURSE

HOW TO TURN OTHER PEOPLE'S TRASH INTO MONEY, PUBLICITY, AND POWER

BY JOHN HOFFMAN

PALADIN PRESS · BOULDER, COLORADO

Dumpster Diving: The Advanced Course
How to Turn Other People's Trash into Money, Publicity, and Power
by John Hoffman

ISBN 1-58160-369-X
Printed in the United States of America

Published by Paladin Press, a division of
Paladin Enterprises, Inc.
Gunbarrel Tech Center
7077 Winchester Circle
Boulder, Colorado 80301 USA
+1.303.443.7250

Direct inquiries and/or orders to the above address.

Visit our Web site at www.paladin-press.com

TABLE OF CONTENTS

CHAPTER 1

Pushing Your Luck

When you're having a great night of dumpster diving . . . when luck turns your way and the reeking decay of the cold, hard universe yields delightful diamonds amid the common and all-too-expected turds, YOU DON'T STOP! YOU GOTTA KEEP ON DIVING! FILL UP THE BACK OF THE TRUCK AND TIE SOME TO THE ROOF, BABY! WIN ALLLLLLL THE CHIPS! PICK UP THAT GIRL AND HER TWIN SISTER, TOO!

Push your luck until it almost breaks, until it dips down and bottoms out amid the Mysterious Brown Black Goo at the Base of Existence, then limp home to count and clean up your blessings, because the next day might not be so lucky, nor the day after that, so it's a damned good thing you did three days of diving in one red-hot night, a night you'll be talking about years from now, like my friend and fan Mark D. in Texas, who pretty much makes a living off discarded computers, including *Pentiums,* for god's sake, and Craig S. in Masontown of the Keystone State, whose diving is so total, so all-encompassing, that the descriptions he mails me seem to have tapped some deep psychic source whose echo is eerie coincidence.

Odd, then, it took me seven years to write a sequel to *The Art and Science of Dumpster Diving,* one of the most successful, influential, and well-publicized "cult" books ever written.

That book took me to Hollywood four times . . . and I can hardly wait to tell you about it, so I can take YOU along with, maybe even leave you there. That book—the one "They" said couldn't exist, because anybody who would WANT a book about dumpster diving would be too cheap to BUY a book about dumpster diving, so consequently what publisher would ever PUBLISH a book about dumpster diving—THAT BOOK roared its existence across the whole world in more than a hundred radio interviews, scores of newspaper and magazine articles, including *Playboy, USA Today* . . . dude, the whole *perception* of who dives and why, the very *term* "Dumpster diving" entered the mainstream lexicon as a result of the media firestorm generated by my book in the mid 1990s and, well, the additional efforts of a few ragged pretenders to my throne of Dumpster guruhood, whose names have virtually faded away.

(Gaze upon that corporate capitalized "D for Dumpster," because that's the last time you'll see it in this book.)

Where was I? Oh, yeah, writing another book. Trying to ride the trend I helped to create and cash in yet again. Will this product be as great as the second *Jurassic Park* or the sequel to *Grease*? How does an author "succeed" by the standard measure (book sales) when he's always telling people, "*Don't buy my book; get it in a library!*" or break a window, anti-WTO protest style, and liberate it from corporate clutches.

Seattle, Washington, was a riot. I spent five years there, and that town liked me so much I got "the keys to the city" (which I'm about to pass on to YOU through this book), but the last and best part was during the 1999 uprising against the World Trade Organization, in which dumpster-dived supplies played a key role in the successful occupation of the Westin Hotel entrance, where important WTO Emissaries of Pure, Undiluted Evil were holed up. You can see my picture in a *Seattle Times* for that crazy week, using an orange traffic cone as a makeshift bullhorn.

But right now I want to tell you about another, much smaller riot in September of 1994, which will be known forever amongst yer radical activist types as the Broadway Police Riot. Here it was I mentally coined a phrase which has yet to be uttered from the lips of my slightly-more-radical-than-me main character in my as yet unpublished novel:

"The best way to recycle a bottle is to throw it at a cop."

So, yeah, we were marching against police brutality, and—as though to underscore our point—they charged us with batons and pepper spray in full view of a citizen named Mike Crow, who was toting a video camera, scavenging like the bird whose name he bears for exciting footage that he could sell to television stations.

I'm in my mid-30s, and supposedly that's still pretty young, and that day in 1994 I was in my late 20s, but something about the guts, the sheer bravado of a sixteen-year-old who doesn't yet truly comprehend her mortality . . . the fact she could die, she could get casually bashed in the brain with a police baton and spend the rest of her precious life in a softly padded home for incompletely assembled human beings . . . it brings tears to my eyes, the glorious guts of our beautiful young people, and it makes me feel like an old man who, at my tender age, has already seen too much.

(I'm getting to my point. My point about sequels and how they are sometimes not a continuation but a missing piece which must serve not as something INCOMPLETE, but the whole damned thing, because that's all we've got, boy, and who decided these things?)

Confession: I may not be the best dumpster diver in the world. I've found a diamond ring, yes, but I've never found a Pentium computer, a live baby, or missing gold-plated Oscars. I supplement my income with dumpster goodies, yes, but I have a college degree (*magna cum laude*) and I make decent money in a "straight" job.

I might only be the best dumpster *writer and teacher* in the world, or the most shamelessly self-promoting dumpster diver in the world, but here you are grasping my sequel in your paws, though odds are only 50/50 that you read my FIRST book, or maybe you'll go back and read it after THIS ONE, which is the SECOND part, and have to switch them around mentally—or you might have to get by with nothing but the sequel and extrapolate all that went before, or try to remember it from seven years ago, but lucky you!

That's how a dumpster diver thinks, anyway.

My point is that I'm the guru you got, and we'll both have to make do, you with my incessant dumpster demagoguery, and me with your silence, which I take for hearty approval of what I'm saying, but not knowing your name and not really being sure we've met before, but here we've run into each other.

It must be destiny.

So . . . *Kids these days!* The innocent bravado of the young who have never before been roughed up by a police officer, never watched their newborn son struggling in an incubator, kids who stand up to a line of helmeted storm troopers with little hand-lettered signs made from cardboard somebody threw away, a corporate logo on the inward side and STOP POLICE BRUTALITY on the outward.

And yet the very handfuls of crap, this NOTHINGNESS which we silver spoonless ones grasp in our calloused paws, even the very cracks of the sidewalk are filled with galactic potential. Because THIS is what happened in September of 1994, and I saw it happen. I witnessed a sixteen-year-old girl with mousy blonde hair transform before my eyes into the Goddess of Liberty—almost too beautiful to look upon—and she is only one goddess you will meet in this book, or perhaps only one aspect of Her.

The storm troopers were coming for us. They were arresting people, dragging them over broken glass. They arrested one guy for walking out of a coffee shop to see what was going on in the street. They arrested his wife for coming to the aid of her husband, catching them in the machinery of their legal system for months and months, grinding away at them. That terrible night, Mike "Video Guy" Crow was on the scene, and—spontaneously—the young people who had organized and led the march turned toward Mike's camera—with police visible in the background, moving in at that very moment—and began making speeches to Mike's camera, as though it were connected to the whole world. (Which it WOULD be, later, with some of Mike's footage making the CNN news loop, but what FAITH those children had at that horrific moment, as they began to speak from the heart . . . .)

"Hey! Is that camera working?" asks the mousy blonde girl, who has a Spanish accent. Her eyes are brilliant diamonds. In that very moment you can see her agile mind grasping the POTENTIAL of Mike's camera, which is all they've got to work with at the moment, since they've run out of 40-ounce bottles—

"They just hit my friend who was pregnant!" says a dark-eyed girl in a black beret. "They just BEAT HER DOWN!"

"They told me that for standing up for our justice, that we would go to jail," says the mousy blonde, with odd phrasing, but she's getting her point across, smacking the side of her hand into her palm for emphasis. "What happened to the FREEDOM OF SPEECH?!"

"Freedom of speech!" agrees another young girl, with long and surreal facial features like an El Greco painting. "Freedom to assemble!"

"This is fascism!" shouts an angry young man. "This is not freedom, this is fascism!"

"We can't even walk down the street without getting beat up!" cries the El Greco girl.

Mike stops his camera at some point, trying to save his lone cassette and dying battery for the scenes of brutality he is sure will be repeated. When Mike goes live again, we see the group of young speechifiers boldly approaching the police line, determined to reclaim their street, to assert the constitutional right to assemble. We can't see the leading girl—the pregnant girl—go down, but at some point in the mayhem we see that she is down. Her friend, Mousy Blonde, jumps to her aid. In full view of Mike's camera, filling the frame, three police smash Mousy down with their extra long batons, all hitting her in the same instant; one from behind, one in front, one just catching her in the face.

A gang bang!

Mike knows in that instant he has captured something incredible, but for a moment, he alone holds its chaotic butterfly potential. Soon all of Seattle would see the dramatic riot footage—over and over, frame by frame—the world would see those images and the prophesy shouted in the streets ("THE WHOLE WORLD IS WATCHING!") would be fulfilled. Scores of languages would comment upon the immortalized images with the polyglot tongues of angels but—at THAT moment—Mike held the footage in his grubby hands, all its amazing potential, fresh, just scavenged from a crap pile of a billion relatively worthless images.

The girl gets WHAMMED on television—and, by the flickering light, excited and sleepless youth are pasting together a flyer, announcing there will be ANOTHER march, the same route, the very next night . . . .

The girl gets WHAMMED on television—and the station gets reports of the planned march (somebody faxed them a flyer) and thousands rush to the scene, a televised feedback loop, the image feeding more images into itself . . . .

The girl gets WHAMMED on television—and three days of marches follow, with thousands in the streets, and helicopters circling above, broadcasting everything. And when the kids get sick of marching they occupy part of the campus and establish a tent city and write up a manifesto, and still the girl is on television, only now she is making speeches, and this time it's not a pathetic little cardboard sign and one lone freelancer with a video camera and dying battery. This time the cameras are live, connected to the world, circling overhead, and authoritative, articulate media voices are saying IT'S ALL ABOUT THE ISSUE OF POLICE BRUTALITY, NOT JUST LAST NIGHT BUT BEFORE LAST NIGHT. THAT'S WHAT THE MARCHERS WERE SAYING, AND THE POLICE DENY THERE WAS BRUTALITY BUT . . . WELL, THE IMAGES OF LAST NIGHT SPEAK FOR THEMSELVES. (Roll tape. The girls gets WHAM WHAM WHAMMED, then cut to live footage of thousands of bodies, thronging in the streets, hanging upon her every word . . . .)

From a handful of *worthless crap*—Mike's dying battery and lone cassette, a few cardboard signs and flyers liberated from Kinko's, a handful of kids with a crazy plan—a media firestorm arose, and its heat swept the world.

The moderate elements rush in. They propose a "youth center" for kids on Broadway. Funding is made available, and an emerald city of sorts arises from the echo of broken bottles.

I look in the eyes of that young woman, that goddess of the streets, and I see the flicker of enlightenment. She has had her revelation amid the flames of a massive media firestorm, the whole world set ablaze by a spark from one of her own thoughts. Now she is on the other side of that line, crazy, visionary thought she dared to follow. In an *instant* she perceived the angle of the camera, her fallen friend, the police batons, and she knew . . . if she just crossed that line, risked everything, suffered for the sake of art and beauty and civilization . . . anything was possible. So it was that, for a moment, and forever, she was not just herself but the incarnation of the Goddess of Liberty, possessing the body of a sixteen-year-old girl.

I mean *woman.*

The occupation of the campus, the tent city, the free supplies that appeared from nowhere . . . a little civilization emerged that lasted but a moment. The most accurate accounts of our revolutionary struggle in the streets were told in unstable little papers that stain the hands, but the words printed with cheap ink became history, and to this very moment they have not quite gone away . . . .

During the September 1994 occupation of the SCCC campus, a clerk at the Chevron station across the street was cool, kicking down free fountain drinks for the young radicals. A pay phone at that same gas station became our "media center" where communiqués were issued.

Where do all the little disconnected pieces come from? Which become so important and resonate forever? The parachute harness that gets transformed into a rudimentary knife? The apple thrown from the truck in a desperate, spontaneous act of caring amid the screams and the gunfire? When you dive into dumpsters, you will find items that become part of your personal civilization the life force in the cells of your children, a portion of ancestral memory passed down for generations, pieces of good fortune that can prove turning points in your whole existence . . . and who *decides* these things?

Who decided that, in 1995, I would find a dumpster full of books about getting jobs on an Alaskan fishing boat, send one to my brother, Slash, and he would be inspired to seek out such a job? And while he was gone, our mother would take in a runaway girl and her sister, allowing them to live in Slash's deluxe converted 1950-era chicken coop, and when the girl was, well, legal enough to get married, Slash would marry her?

How can the things which are so *important* be so random and chaotic? And what is more *random and chaotic* than broken pieces fished out of the refuse and fed back into society?

Here is a broken piece for you: a sequel to a book . . . that wasn't supposed to be written in the first place . . . the "other half" of something missing at the moment, a hunk of weird destiny, of eternal consciousness . . . a lever for you to move the world.

So there I was upon the occupied campus, where we expected the police to mass up and rush us at any moment—sleep deprived, emotionally exhausted—but using all my indigenous urban hunter-gatherer skills to pull free energy into the riotous media feedback loop, thinking how IF WE COULD TAKE OVER CITY HALL, AND THE POTENTIAL WAS RIGHT HERE, RIGHT IN THESE BOXES OF MOTHBALLED SWEATERS THAT AN OLD LADY DROPPED OFF AND THAT PILE OF TWO-DAY-OLD BREAD, AND I COULD FIGURE THIS OUT, BECAUSE I'M THE FAMOUS TEE-VEE DUMPSTER GURU WHO ACHIEVED BODHISATTVA ENLIGHTENMENT UPON A BALCONY IN BEVERLY HILLS—and right then a mysterious grizzled stranger interrupted my thoughts by offering me a cup of steaming hot coffee.

I don't drink coffee (nor alcohol, actually), but he kept insisting. And it was a "tribal situation," where nuances and gestures of hospitality are important, and he kept INSISTING, so I took the coffee and drank from a hateful Styrofoam cup, which he warned me to recycle, and I said, "Mmmmmmm."

"Is it GOOD?" he asked, but it wasn't REALLY a question.

"Oh, yeah, that's great coffee!" I agreed, to please him.

And I kept trying to figure out how to put together all the broken, discarded pieces—but my mind kept racing wildly ahead, making connections I hadn't seen before and hadn't been looking for—then suddenly, THE LINES IN MY HAND got very interesting, more interesting than they'd EVER been in my life, though years later I felt that way about the hands of my little newborn son . . . and when I looked up I saw three angels, coming toward me, manifesting in the form of young men who lived on the streets, who lived out of dumpsters, who survived that way even to a greater degree than I did, but they certainly weren't famous for it. And at that moment I had a revelation, about how "sensitive dependence on initial conditions effects the galactic destiny of gonads, such that every little thing we do or say impacts what civilizations will spring up in which star systems, where interplanetary wars will be fought or peace achieved . . . ."

Yes, at some point I realized I had been dosed with LSD.

But what was I supposed to do? Turn to somebody and ask for help? Or experience it and see where it might lead . . . .

Push my luck?

God knows where this unexpected, weird trippy trip back into the dumpsters might end up. The kind of books published by Paladin Press get more dangerous to sell every day, yet Paladin endures.

Whatever happened to our freedom of speech?

Welcome . . . back?

Grab your gonads. LET'S PUSH OUR LUCK!

CHAPTER 2

The Master Key to Successful Diving

OK, I can see that this little group outing has already turned into chaos. At first I couldn't hear you, but if I concentrate upon that corner of myself that got dosed with acid, I can actually make out your voices.

Some of you are excited and eager, already quite familiar with dumpster-diving concepts through experience and my first book. But some people in this class . . . well, they are stuck with *just this book* on a long bus ride. They are Madison Avenue trend spotters slumming the tenderloin of the publishing world, looking for opportunities to cash in, and some folks in the back are asking questions that lead me to believe they are police informants. But the real problem here—besides the possibility law enforcement could swoop down and halt this little enterprise before we even get to the end of the book—is that we've got novices in the same class with folks who have almost advanced to "master" or "guru" level themselves. The novices aren't even convinced there is any *point* in diving dumpsters.

Pardon me while I talk to the novices for a moment. Bear with me, those of you who already know me.

Do you despise banks? *Huh?* Well, let me tell you about using dumpster-dived documents to mess with banks. And when I say "mess," I mean, like, put them on NBC prime time, during Nielsen "sweep week," as part of an ongoing KING-5 TV investigation.

What? You say that all the good dumpsters seem to be locked?

Oh, now I can hear more people saying that. Even the most advanced divers are pointing out the frustration of locked dumpsters, which seem to be increasing, partially as direct "negative feedback" to the dumpster-diving trend.

Well, that's why I'm the guru. That's why I get paid the big bucks. Besides, this will be a good place to cover "master keys to dumpsters," right in the beginning, in case something happens, one of the police informants pulls the fire alarm and the class has to evacuate or something. We're only on page what? Eight?

Mentally stick out your hand. I can see some of you are getting excited as you hear a sound coming from this rather large, imaginary canvas bank bag that sounds like *old coins* or something. Here, my friend, my revolutionary comrade, my precious child, here's a key for you, and for you, and for YOU. (Somebody help me pass out the rest of these keys . . . yes, you, the Paramount Studios tour guide . . . YOU look trustworthy . . . .)

What? You can't *see* the key in your hand? You're getting pissed off because you're saying it's a *pretend* key? What if I told you that keys which unlock thousands of dumpsters are easily obtained almost *anywhere?*

And I'm going to tell you *where* to go get one? Easily? TODAY, this very hour?

OK, now this frustrated fellow here is beginning to see the point of the "pretend" key. And that guy in the back, with the crew cut, is getting really nervous.

Another reason to tell you this, right now, is because I discovered it *after* my first book was published, and it's probably the most important thing that got left out. Return with me to Seattle, Washington. (Better pack a gas mask!) When me and my wife, Tina, blew into town in May 1994, we discovered a town where most of the good dumpsters were locked, a town that stood in stark contrast to El Paso, Texas, where we'd been living, where the police will practically hold the lid for you while you rummage and, needless to say, there aren't a lot of padlocks.

It was very frustrating.

After writing *The Art and Science of Dumpster Diving,* I also became more intimate, more familiar with the Dumpster Goddess. It's not like I've become a pagan or something, even though I've noticed a direct correlation between offering a "sacrifice" of dumpster-dived bread to the pigeons, crows, sparrows, and sea gulls before diving, and having good luck during *that* dive. Mostly I think of the Dumpster Goddess as Lady Luck.

Come on! I think. *Seven come eleven, got to be something good behind this desert convenience store oh–!*

SNAKE EYES!!!!

So I was frustrated. And I cried out to the Dumpster Goddess. Even though I don't "believe" in her or "worship" her, I cried out to the Goddess in my moment of need.

All these damned locked dumpsters! Should I just go through the alleys and cut all the locks? Mother may I? I could become a legend in my own crap mine. Johnny, JOHNNY Bolt Cutter . . .

And She gave me an answer. The key, if you will, to advanced dumpster diving.

It was the very next night, and I was going through this sort of mental process: *Bolt cutters. Must get bolt cutters. Buy them? Not going to buy them new, THAT'S for sure. I could pick up a good used pair at Hardwick's. Hmmm. But I wonder if I could borrow a pair from Bicycle Bill? But what if the cops pull through the alley and I have to ditch Bill's bolt cutters? Well, THEN I could replace them with a used pair from Hardwick's. Hmmmm . . . another fucking lock, and if I cup my flashlight just so, between my hands to shield the light from prying eyes, I can see last month's issue of HIGH TIMES right through the space between the lids! Hmmmmm. I don't do drugs but I sure love to read about hydroponic advances! I AM SO FRUSTRATED! Wonder if I can pry up the corner of this lid . . . . ?*

Down the alley a little further, I advanced upon a dumpster for a sort of yuppie pottery and furniture place where, once or twice, I had found soothing rural images beneath broken glass, surrounded by perfectly good frames . . . but only when they left their dumpster unlocked. With so many dumpsters in so many alleys, it often made sense to ignore the troublesome locked dumpsters and simply move on to a dumpster that *wasn't* locked. But I can usually spot an unsecured lock from far away, a dumpster that only *looks* locked, and on this particular night I was focusing upon that yuppie store lock from afar down the alley, thinking like this . . . .

Locked or unlocked? It looks wrong. Something about the lock looks wrong. What is THAT sticking out of me . . . ? Oh, geez, is it what I THINK it is?

In that dark alley I laughed hysterically. One of the employees of the yuppie store had left the key sticking out of the lock. Triumphantly, I yanked it out. I rammed it back in. I snapped the lock open. Inside were the usual shards of pottery, broken frames, evil Styrofoam peanuts.

My very own locked dumpster! I thought. *I have a key to this dumpster now, and it's like my own reserve! Only I will be allowed to hunt here! Excellent!*

I pictured how many keys I would need to carry around to open all the dumpsters in Seattle, and how I would look like a mad key collector. But there's something about *keys* that lights my imagination with possibilities. Once, I found a key that bore the words "Change-o-matic." The key, I think, to a money machine, so I've childishly added it to my key ring, thinking one day I will come upon the money machine. I've even been known to tell people's fortunes with keys. As digital cards gradually replace familiar keys, they will become collector's items, like arrowheads. In any case, purely to amuse myself, I added the key to my ring. And the next night I was in that dumpster again, doing my Thurston Howell the Third imitation. ("Ahhhh, my own private stock, Lovie!")

Only something was *wrong*. This didn't look like the garbage from the yuppie store. I realized that, in the dark, I had opened the wrong dumpster . . . *but the key worked anyhow.*

Wait a minute! I thought. *If it opens the first dumpster, and the dumpster right next to it from a different store, I wonder if it opens* more *dumpsters in this alley?*

At that point I was thinking in terms of the garbage truck route, thinking to myself that maybe all the trucks on one particular *route* might use the same key. This key, I figured, might open up most of the dumpsters in my neighborhood. Of course, as I quickly found out—dashing down the alley, madly sticking my key into every lock that presented itself—the master key only opened locks from a particular company. But since that amounted to approximately 80 percent of the locks in my neighborhood—and most of the rest of the locked dumpsters were "recycle bins" designed to hold cardboard—I was ecstatic to discover I could unlock my whole neighborhood, all the way from the park in the north (where homeless youth often encamped) to the edge of the water in the south near the university, all the way from the very spot in the very alley where disgraced former garbage gleaner Ted Bundy once abducted a woman named Georgeann Hawkins (and now, decades later, I could *still* find photos of a sorority sister pajama party in that very alley) to the gas stations and strip malls near the highway. My whole *neighborhood!*

I had to tell somebody. *I had to keep it a secret.* I had to figure out how to copy this key. *I had to hide this key!* I had to give this key to *everybody!*

I rushed over to see Bicycle Bill at the Center for Urban Ecology. Bill, the nearly seven-foot-tall genius giant who once waved a crowbar in the face of Seattle Mayor Norm Rice during the course of a building occupation. Bill, who concerned himself all day long with the advantages of light rail versus monorail and composting pilot projects and the works of Buckminster Fuller.

Bill, who hated those locked dumpsters as much as I did!

"We've got to be careful who WE give this key to," Bill said.

"Why?" I asked. "Let's open all the damned dumpsters! Let's give the key to everybody!"

"The carting company might change the locks," Bill argued. "They could come down on us hard. This is criminal shit. You're actually opening a lock with a key that's not yours."

"I haven't made copies *yet!*" I reminded Bill, slapping my hand down over the small, precious key, which he was gazing upon the way a cloistered nun might gaze in adoration upon the Eucharist.

"This could be a *five-year key!*" Bill said. "If you give it to just anybody, if you give it to some *jerks* who will throw crap everywhere, it will mess up everything. The whole thing will become *fractalized!* The problem is that anybody you give a key to—"

"Anybody I give a key to," I finished, "can duplicate that key and give it to somebody else, and all of *those* people—"

"Sensitive dependence upon initial conditions," Bill muttered. "It would replicate exponentially, creating the same fractal pattern as the branches of a tree, or an epidemic."

"Yeah, yeah," I muttered, waving him off. I just wanted to figure out how many corporations I could *fuck* in the shortest span of time possible. I wanted to unleash all the power of the key at once, and here Bicycle Bill—he of the crowbar in the mayor's trembling face—was urging me to be careful, to plan my steps with sensitivity toward the fuzzy buzzwordy principles of "chaos theory," which, he warned me, one does not dare attempt to harness.

"The butterfly flaps its wings and creates a storm on another continent," he prophesied. "The storm comes and *kills* the butterfly!"

"Why doesn't the butterfly *just fucking move* if it knows the storm it created is coming back down the pipe?" I asked, spawning Hoffman's Theory of Chaotic Butterfly Self-Preservation. "Why doesn't the butterfly just get out of Dodge?"

"Hmmmmmm," Bill muttered, gazing at a map of Eugene, Oregon, he had taped to his wall. "You could be on to something there."

He smacked his fist down upon the table.

"You don't even know how *powerful* The Key might be!" Bill continued, scolding. "We *assume* it's a key for this neighborhood, for [here insert the name of the refuse company], but you don't really know the *parameters* of The Key. Have you encountered any locks from that company that you *couldn't* unlock?"

"No," I shrugged. "But I haven't gone beyond 15th Street, up into Bundy Country. I haven't crossed the highway or gone beyond Ravenna Park."

"What if this key opens the whole city?" Bill asked, rocking back and forth, with that scary *visionary* look he would get when talking about incorporating hydroponics right into the walls and floors and ceilings of future high-rise buildings, all the while *himself* living in blatant violation of ten city codes inside an old wooden office building without central heating or flooring in the root cellar, spinning out "sustainable urban ecology pilot projects" while *himself* living off free Alaska salmon at the food bank, haunting the same dumpsters as me.

"That would be crazy," I answered, but I had been wondering about the possibility of a key to the city myself. "Would they be so *careless* as to use *one lock for everything?"*

We looked at each other.

"Of course they would!" we said, together.

"They're a hierarchical system," Bill noted.

"They assume people who want to get into dumpsters are *morons* or something," I nodded. "It wouldn't even *occur* to them that somebody would discover the weakness in their simple, cost-effective system and systematically exploit it."

"You've got to figure out the *parameters* of this thing before you unleash it," Bill said, waving his finger in my face. "You have to be *responsible."*

"I agree," I said. "I have to know exactly how much chaos this thing can unleash *before* I unleash it. So I can unleash it in the best way possible. This key might very well open up the entire city of Seattle."

"Everett, too," Bill said. "This same company is up there in Everett, Washington."

"This is going to be a pain in the ass," I muttered. "I'll have to go east and west and south and north . . . ."

"No, just go downtown," Bill shrugged. "Believe me, if it opens dumpsters downtown, it opens everything else. Downtown *über alles.* Don't even get me started."

He meant: *Don't even get me started about the million-dollar parking garages built with HUD money, the low-income housing destroyed to make room for a decorative topiary next to a publicly subsidized convention center . . . in short, the way a prosperous downtown Seattle came at the expense of everything and everybody else, a pattern which pretty much replicates itself across the country and the world. And so, as an easy rule of thumb, a master dumpster key that works on downtown dumpsters probably works everywhere.*

What a strange little journey this is, I thought, as I rode mass transit downtown to "determine the parameters of the dumpster key" by applying Bicycle Bill's "downtown *über alles*" theorem. Since I had business at my lawyer's office (I had been arrested for sitting in front of a bulldozer to defend homeless people living in a "greenbelt" near the highway), I figured that an alley in Pioneer Square (near the purported original "Skid Row," where once the logs were skidded to the water and the broken men would congregate, though other and lesser cities lay claim to the origins of the phrase) would work as well as any for this historic test.

Entry to the alley for non-business purposes was *verboten* by yet another anal-retentive city ordinance a la city attorney Mark "of the Beast" Sidran. The alley was full of winos and junkies, however, plus one locked dumpster overflowing with crap, but its lock was still engaged, the chain stretched to its full length. I would have *never* dived this particular dumpster in that particular place unless I had reason to believe somebody had ditched a bag of money from a bank robbery in it. My interest was, shall we say, purely academic.

Snap!

It came right open.

I stood there, stunned, and a little bit frightened.

I should have known, I thought. I am the master. I've dived Hollywood, I've found a diamond ring, I'm the world's foremost authority on this subject, *and yet I didn't know refuse companies might be in the habit of using identically keyed locks for an entire city.*

Which made this key, oddly enough, almost valueless and besides the point. True, I was holding a very special key that unlocked most of a major metropolitan area. I could give that key and all its power to anybody I chose . . . .

But anybody could get this key almost anywhere. Every business, naturally, needed a copy of the key to unlock its dumpster. Every busboy and floor sweeper and developmentally disabled dishwasher person working his way up to middle manager had access to this same key, no doubt affixed to a variety of difficult-to-misplace objects hanging upon rusty hooks: dustpans, hubcaps, even a rubber chicken. What was really valuable, however, was *the knowledge* of the key.

I remembered when I was an apartment manager and we rented some nice young man an apartment. But we gave him the wrong key. We gave him a *master* key by mistake. For days he was going in and out of his apartment, unaware he could have unlocked *all of his neighbor's apartments* and made off with their stereos. Could have stood there quietly as they hummed, serenely, chopping veggies in the nude. Could have made a mistake, entered the wrong apartment, got himself shot . . . . .

Do other cities use a "universal key" system like Seattle?

I have it on good authority that some of them do, including but not limited to a major city in Texas. In fact, I understand that—shortly after this "master key system" was revealed in my 1996 video *The Ultimate Dive*—keys to certain cities started getting traded at punk rock concerts. (I can't prove a causal connection, however, and it is possible, even likely, that other individuals have known about and made use of universal dumpster keys at various places and points in time.)

So, logically, you must first establish whether the major refuse companies in your city—if they are locking their dumpsters and this is troublesome for you—are using a universal key system and then, secondly, you have to get one of the keys.

First, let me kick myself MORE. In my previous book, *The Art and Science of Dumpster Diving* (hereafter referred to as *TAandSoDD*), I illustrated a point about locked dumpsters with the story of "Mrs. Spooner," who had a beef against a particular organization, and she figured out how to get into their dumpster by simply buying padlocks until she found one with a key that matched. In the next chapter we'll be talking a lot about how to humiliate corporations, politicians, and governments with dumpster-dived documents and advanced dumpster-based media manipulation. For now, I have to finish kicking myself . . . arg!

There.

We need to move beyond single dumpsters and single locks. Having a universal key is a great advantage and, to some degree, distinguishes novices from masters. First, look closely at the dumpsters in your city. (Actually, look closely at EVERYTHING, but let's start with dumpsters.) Note their color, style, and distinguishing marks of corporate control. Note their warning stickers, and remember, we are not "playing" in the damned dumpster. And leaning into a dumpster is not "entering" a dumpster, but tell it to the judge.

Yes, I digress. I always digress. It's my nature. I'm a scavenger. I see a direction that looks good and I go that way, then I find my way back.

Until I tried to figure out that wonderful key in Seattle, I hardly gave a damn about who controlled which dumpsters, their color or style. I might take note of some unusual feature ("Ah! A locked box with an open top two-holer feature. Haven't seen that since El Paso!"), but obviously I'm more concerned with the contents than the packaging. Again, it is my very nature to care little about packaging and focus more on the goodies inside.

But once I fixed my attention on the dumpsters of my fair and foul Emerald City, trying to master the "magic key," I learned to take even MORE joy in the act of diving. There are things you can take for granted . . . like the love of your wife. BUT IN ANY CASE!!!!! Don't be fooled by the colors of dumpsters. You might find that brown is "standard refuse" and white is for recycling paper and cardboard, but it's the same company, but it's different locks. You may find that a company uses an older style of dumpster *and* a newer style, and these styles have different colors, but the same locks. As the powers-that-be react to widespread knowledge of dumpster keys, you may find one lock being phased out and another phased in. It's fairly easy to tell if more than one system is in place if you're encountering more than one kind of lock.

And actually, it's easier to figure out the system AFTER you have the key and you discover, by trial and error, what it unlocks. Yet BEFORE going after a key, you will want to scout around and see what sort of system appears to exist. It's kind of like robbing a bank. You won't really know how it's going to go down until you're holding that piece of metal in your hand. But you will certainly want to get a rough idea of where the guards are and what their routine appears to be.

So, let's assume you've looked around and you see a lot of locked dumpsters. Well, unless you get really lucky and see one with a key sticking right out of the lock, you will have to obtain one from a key source.

Remember, WHOLE MOBS OF PEOPLE HAVE ACCESS TO THAT KEY. It's likely that you know one of these people. Your roomie, Kelly, who works in a fast-food place to pay for college—who has a million schemes and dreams for fame and fortune—he might have instant access to that key. Or maybe Paula, who plays in a band and stocks shelves for a hateful day job, her tongue constantly probing the empty hole of her corporate-disapproved lip piercing . . . maybe she has the key. Maybe she wears it with her dog tags. You can wear the key in your piercing, too, a "Prince Albert" that lets you *into* the can.

Yeah, OK, that was going too far, even for me.

But hell, it might be the key in the very place where YOU are working. The point is that these keys are floating around all over the place, and you might be—what?—only one or two degrees of separation away.

Obviously, you don't want to get caught. So you don't want to ask somebody who is an *asshole* to duplicate the key for you just because they're the only person you know with access. Likewise, visiting a friend at lunch hour, in the steamy back room of the establishment, striking up an awkward conversation and then trying to swipe the key probably isn't a good idea, either. And, obviously, the garbagemen have access to these keys and they might be in your circle of friends, but those keys are even more closely linked to job security for them than, say, a waitress.

If you're *totally* committed to dumpster diving and locked dumpsters are cramping your style, one option might be to get employment . . . for all of one day. Just long enough to swipe the key and make a copy during your lunch hour. (Even if you're caught walking off with it, you just say, "Oh, hell, I left it in my pocket.") Just don't walk back into the business establishment with your new key and a receipt from the locksmith. And, well, be really careful if you're going to dive the dumpster at that very location.

Since writing *TAandSoDD*, I've met and dived with a number of wonderful fans, including Chad, the husband of Robin, in Washington state. I find that my fans are more daring and take chances that I wouldn't dream of taking, like hotly debating store managers over refuse disposal. Suffice to say, in most cases I probably would avoid diving the dumpster of the place where I had been employed unless I had a very specific and important reason for doing so . . . like I wanted to "get something" on my old employer. And, as Michael Moore once pointed out, the time to "get something" on corporate employers is when you've got access to their files, certainly not when you're in the parking lot wondering where you will work next.

Suffice to say, copying a key from a particular location and then later showing up at that location with the key is not a wise policy . . . but I'm flexible enough to rule out nothing. Like I said in my video: *It's not like what I'm saying is dogma or something. It's not like you HAVE to do it this way.*

Oddly enough, I don't think many people learned to dumpster dive from my book. Most of the people who write me were already dumpster divers and simply wanted more tips, or for whatever reason they enjoyed reading about this activity. Lots of people bought the book as a gift for somebody they knew who was a dumpster diver. (Whether they always meant their gift in a purely affectionate way is open to speculation, but the royalties from those book purchases spent just as well.)

Moving on: another increasingly attractive option is to use the Internet to find somebody who has what you need. With all the publicity about cyber drug deals and kiddie sex, it's unlikely the police will be setting up seekers of dumpster keys any time in the near future.

So let's assume you've managed to obtain a key or keys some way, somehow. The first thing you need to do, before you unlock a single dumpster, is to make a copy. There appears to be little likelihood, at least for now, that dumpster keys will be marked DO NOT COPY. However, I don't doubt such a key might exist. Perhaps there are dumpsters at sensitive government facilities, and those keys, like practically any key in such a facility, are big, thick mothers stamped with warnings. If another "cycle of social upheaval" takes place, we may see such widespread modifications, the way cash registers obtained sloped angles after Abbie Hoffman wrote about a certain trick clerks used to steal money, keeping track of how much money they stole by placing pennies upon the register.

Hey, let's face it—the police informants are reading this at the same time as the rest of you. In fact, my first dumpster book was recently cited in an FBI Law Enforcement Bulletin about "identity theft." (Hi, boys! It's me again, spreading a little creative chaos, keeping you in business. You should THANK me. Mail me some cool law enforcement stickers. I LIKE THOSE, and I have a special filing cabinet full of nasty internal affairs documents and "bad cop" scandals that I like to decorate with police bric-a-brac.)

But, as we proved in Seattle, and as Assistant Police Chief Ed Joiner admitted after the uprising against the WTO, authoritarian systems can't "redeploy their forces" as quickly or effectively as we can. If we will all adhere to Cardinal Rule Number One of dumpster diving: *Don't make a big mess!* If we will leave a dumpster in the same condition we found it, only better (better is emptier) and relock that which was locked, and leave unlocked that which was unlocked . . . well, then, like Bicycle Bill said, "This might be a five-year key."

Yes, the system will evolve. The day will come when we see magnetic card locks on dumpsters . . . and the day will come when master codes for such locks can be hacked and downloaded off wireless Internet systems.

Such a wildly evolving universe we live and dive in! Some of the novices might be surprised to learn that the master diver sits around, munching upon his asparagus and mushroom dumpster delight liberated from the local Leevers dumpster, and speculates about how, one day, some human beings will not only avoid eating animals but will avoid brutalizing innocent plants as well by incorporating photosynthesis directly into our bodies through genetic engineering.

We will be like angels, living from light. Also, ultimately, I believe we will have real fairness and equality when women are entirely liberated from the burden of bearing children. We will use artificial wombs, not like the sterile "decanters" in Brave New World, but—through genetic engineering, the same dangerous miracle that put a gene from the North Atlantic cod into a tomato—we will literally grow human beings like fruit upon trees. In other words, babies really *will* come from the cabbage patch, and storks might be involved as well.

How exciting, to see the world evolving every day toward the stars and the angels . . . and to play a role in that evolution by foiling a widespread system of locked dumpsters, which will ultimately give way to another system, which will need to be foiled by, perhaps, another master. How *exciting* to obtain secrets about powerful systems from poorly secured refuse disposal sites and laughingly release that knowledge into the information stream, forcing these powerful systems to address and fix their many flaws!

I've been in this dumpster game for a long time, and I'll tell you this: the people you meet who dive, they are the most creative, the most visionary, the most *connected* to the flows and nuances of life. I'm proud to be associated with all of you.

Now, back to our key here.

If you've only got *one* of something, and it's rare and hard to obtain, but it can *replicate* itself . . . well, the first thing you want to do is make a copy, naturally. That's why those scientists didn't *eat* the mammoth they dug out of the ice in Siberia. First they're going to clone the damned thing, then later they'll think about mammoth steaks, or miniaturizing them and coloring their fur blue and selling them in pet shops next to the glowing green monkeys.

Make a clone of the key and hide the original away. Always make copies from the ORIGINAL, not from the copy. Everybody knows what happens eventually when you make a copy from a copy. Never risk your original, not even for one little dive on the way to the hardware store. You might not get caught, because you are smart and careful, but what if the lock is gummy and your precious key gets STUCK?

Another reason to have a copy is because a key to a dumpster is incriminating as hell—this is over the bounds of legality, kind of like diving a dumpster behind a fence clearly marked NO TRESPASSING—and if a policeman suddenly pulls into the alley and doesn't dim his lights, if a security guard comes around the corner and breaks into a sprint, you will need to hide that key or, better yet, ditch it in a hurry. And, unlike a busted mop handle with a nail in the end ("dive stick") that you can just drop in the dumpster and walk away from, a key is very specific, very criminally hot, very troublesome if it is connected with you.

However, businesses should pay close attention to the recent case where K-Mart was hit with a $5 million judgment awarded to a lovely couple accused of diving the dumpster and being tackled by an overzealous security guard. There have been other cases in which newspaper editors have railed against the ridiculousness of prosecuting people for "stealing garbage."

If mainstream corporate types are reading these words or, more likely, having them passed on through some intermediate source, here's some free yet very valuable advice: when you go after a dumpster diver, you are *always* the loser. You will spend piles of cash in legal fees, you will risk a public-relations disaster, *and you will deter nobody.* Exterminating mice in your condo with a *jackhammer* is just about as smart. Instead, why not be *clever* and *evolve* by figuring out how to make *money money money* (oooooh, the lovely word!) off something you were going to *throw away?*

Don't keep a dumpster key on your ring of keys. I am guilty of breaking this rule, but then again I am guilty of breaking most of my rules, but the risk is that you would have to yank it off the ring in a hurry, or ditch your keys and replace them, or risk having those keys *connected* with you . . . Just don't do it, OK? Use a piece of string, a thick rubber band, whatever, and keep the key around your wrist so you won't have to put it in your pocket, pull it out, put it in. When you finish a dive, get in the habit of yanking the key up past the level of your sleeve if you're wearing long sleeves. If you're not wearing long sleeves, well, maybe just keep it in your pocket after all. Another alternative is to pin it just inside the front of your shirt—inside a pocket or behind your buttons or snaps—at about the level of a dumpster lock.

Do what feels comfortable, but remember that when you gracefully lean headfirst into a dumpster ("the diving swan position"), whatever is in your pockets might obey the rules of gravity, even if you—like me—habitually break rules, memorizing yet *more* rules for the purpose of breaking them.

One of my most embarrassing moments happened while I was diving a dumpster with an investigative news crew nearby and my ring of keys fell out of my jacket pocket and right into a pile of crap! I am glad I had this moment of mortification all to myself and that I found my keys rather quickly, but I will never forget that jolt of horror and realization. Here I was their "expert" and I had made such a novice mistake! I share that moment with you, however, for purposes of illustration.

Don't lose your dumpster key in a *dumpster.*

Be the master of your keys, conscious and careful of their powers, because in the wrong hands—the hands of a store security guard, for example—your master key could be the master of YOU. As I've warned them just a few short paragraphs ago, corporations will only suffer by going after dumpster divers, but you will be suffering, too, up to and including dealing with fines and probation (but probably not jail time unless you have prior criminal convictions).

THE WEIGHT OF THE KEYS IS INFINITE

I've had many moments of enlightenment and mastered many areas of human activity besides dumpster diving, and once upon a time, in a certain medical setting, I seized a set of "controlled narcotics keys" from a drug-impaired nurse, who then threatened—in her manipulative substance-abuser way—to frame *me* for the theft of Ativan. With Satan speaking through her mouth, trying to get to the root of my innermost fears and doubts, she said: "John, everybody knows how street-smart you are. How you find stuff in people's trash and sell that stuff for money. What if I 'confessed' I was stealing Ativan and you were selling it? But that you're so street-smart you sell drugs without *using* drugs, which is why you never get caught."

Ah, the real-life perils of being a famous dumpster-diving cult author with a nice paying day job.

She (her name was Diane C.) asked if I were "ready" to assume The Responsibility Of The Keys, the weight of every drug in the double-locked cabinet, the burden of every life in the facility.

And I laughed in her face.

"I am the union shop steward!" I answered. "Everywhere I go, I have aspired to positions of tremendous responsibility. You think I'm scared of you? Of the possibility you'll accuse me of something or file some pathetic little lawsuit? I SUE THE *POLICE DEPARTMENT* FOR SHITS AND GIGGLES! I SIT IN FRONT OF BULLDOZERS. I FLY TO HOLLYWOOD. *BEAUTIFUL WOMEN FLOCK TO ME!* I WOULD TAKE THE KEYS TO *HEAVEN AND HELL* FROM THE HAND OF SAINT PETER! AND GOD WOULD TELL ME, 'WELL DONE, MY SON.' SO PICK UP THE PHONE, DIANE. *CALL MANAGEMENT!* CALL THEM RIGHT NOW . . . *IF YOU'VE GOT THE GUTS!"*

In reality, I was chilled to my bone marrow (which I've sold for money to CellPro, Inc, "seeking the dumpster within . . . Ohhhmmm") at the thought of her false accusation in the hands of my enemies. My fear made me weak, but I concealed it, and I was not as weak as her. She crumbled. And I emerged a hero who solved a problem nobody else seemed able to solve. And the awful truth was, I was going after her *vehicle keys,* fully prepared to keep my mouth shut if only Little Miss Too Wasted to Drive, RN, would call a taxi, but I grabbed those *narcotics keys* by mistake, and then I had to play it off on the spot and deal with the problem nobody wanted to deal with.

So if you grab a set of keys, bear in mind you might be taking on a grave and terrible responsibility. It's one thing to find something in a dumpster—missing Oscars,[2] a live baby—that forces you to take a course of moral action. It's quite another thing to explain how you got into a *locked* dumpster. But the problem of explaining matters ("Um, it was unlocked when I got there, behind the fence, <shrug, shrug>") may not be so great as the "problem" of finding that *locked* dumpsters contain many more interesting things than *unlocked* dumpsters. They are, after all, like little sideways rooms . . .

. . . and if something in a room is *important,* isn't that room very often locked? Giving you a key, and telling you how to obtain a key, is easy. Trying to convince people to be *responsible* with the key is not so difficult either. These are, after all, just *words* upon a screen that I will sell for money. The real burden of the keys will be YOURS once you take them.

Once I went to a union training class at a local college, and as we walked behind a science building, we could hear a kitten that seemed to be trapped in a dumpster. I wasn't actually *aware* that I was behind the science building until a member of our group said, "I wouldn't go in there! That's the science building. God knows what you might find in there."

Right. Perhaps I might find a partially disemboweled kitten with electrodes sticking out of its head. The implication of her words being that it was better to ignore the kitten's cry for help and pretend nothing was wrong, because lifting the lid might cause a problem, a situation, a need to act when it was so much easier to walk away. Horrible things are happening in labs every moment; should I make all of those moments *my business?*

I am kind to animals. It's true that my little brother, Slash, currently makes a living driving a truck that picks up dead dogs and cats from the so-called "Humane Societies" all over the country, which are packaged up and sold to science students, further desensitizing them to the plight of animals and assuring a steady supply of animals being dropped off at shelters.

A feedback loop.

Slash also picks up cattle gonads, which are called, inside the company, "cock 'n' balls," but he is, generally, kind to animals too. Once, briefly, Slash tried to be a trapper, emulating the once-upon-a-time lifestyle of our maternal grandfather.

"I had to club a fox," he said to me, tearfully. "It was like killing a puppy. I couldn't stand it. I don't know how anybody with a heart can be a trapper. I'm not going to do it any more, by god. Well, except for muskrats. They're different. I mean, who gives a shit about muskrats?"

He's trying. I assure you, he's trying. And I make up some of the slack for him. Because I went right for that damned meowing dumpster in a heartbeat, heedless of my companion's warning.

If I had found a kitten in there, mangled in the course of a science experiment, I would have made a federal case out of it or, better yet, a *media* case. I'd probably miss the rest of my important training that day, and no doubt the fact I was there for a union event would have affected the relationship between the union and the college. The incident would have been immortalized and remembered in a generation, given a moniker like "The Hardin Hall Electric Kitten Incident."

Am I supposed to *ignore* such a shot at immortality? Walk away from it? Just because it is painful, uncertain, and potentially has its bloody, sickening aspect?

As I lifted the lid, a kitten that had been scavenging *under* the dumpster skittered past my feet.

I was relieved. But I had been prepared to act, to assume awful burdens and responsibilities spontaneously, in a moment.

Bear in mind that even entering an unlocked dumpster can thrust grave responsibility upon your shoulders, but this is so much more so when you are entering a locked dumpster. That is, after all, one of the common symbolic meanings of a key: *responsibility*. But the other symbolic meaning is . . . *treasure!*

Just do the best you can. In a complex system, the outcome of any particular action is almost impossible to predict.

Act in the truth and circumstances of the moment.

AND DO NOT GET CAUGHT WITH THAT KEY.

CHAPTER 3

Advanced Dumpster-Based Media Manipulation

It's a strange and twisted life I lead. Sometimes I have to think to myself. "Well, I better avoid that dumpster. I was there with my personal videographer just last week . . . Better avoid that one, too. They're still upset about the investigative news thing . . . Hmmm, wonder how much I can get for these *News Radio* scripts I found in Hollywood?" (What was I doing in Hollywood *that* particular time? KNBC in Burbank flew me out there for a story about dumpster diving . . . to pump up their ratings . . . so they can sell more damned *products* and create more *garbage.* I had my own personal driver from Medellín, Columbia, who took me to the front of the line waiting to see *The Tonight Show.* "Coming trew," he said. "Guest of NBC.")

How do you manipulate the media with stuff you find in the dumpster and take down entire corporations? By using a little bit of "magic blood" or "fuel" to create a feedback loop, just like those kids in the Broadway Police Riot. It works along the same principles as a shark feeding frenzy or the Dresden firebombing during World War II. You must understand a feedback loop in order to use the stuff you find *effectively,* because there is no doubt you will find nasty, incriminating stuff.

To use the shark analogy, a little blood gets in the water. Sharks smell the blood and rush to the area. Frenzied, they end up biting each other. Which means more blood in the water, which means more frenzied sharks rushing to the area, which means . . . .

So why don't all the sharks in the world eat themselves at once? Well, because butterflies flapping their wings in Peru create tropical storms and sea currents that dissipate the blood. Same thing with a firebombing. Incendiary bombs create a big ball of flame. Needing oxygen, the ball of flame will suck wind toward itself. The wind can be so powerful that it picks up items like paper, leaves, boards, hapless human beings—in other words, fuel, which makes a *bigger* fire, which sucks in *more* fuel . . . .

But it can't and won't last forever.

The media is not all wise and all powerful. But they try to project themselves that way. When I was the editor of a little paper in Minnesota, I would hear people griping about how the media "missed" a big story. For example, there was a huge kiddie porn investigation in one little town, which *some* people caught wind of, and they would call, anonymously, with tantalizing hints, but never enough information for anybody to get a grip on the story and dig the damned thing up.

A lot of people are under the impression the media just "knows" stuff, they somehow learn by magical osmosis where big, hot stories are hiding. Wrong, wrong, wrong! The media get their stories because *somebody initiated contact.* And if you are that somebody, in order to be truly effective you have to be able to pitch the story in a virtually complete form. In other words, do most of the work for the journalist. If you're going to give him a damned *piece* of something, it better be a pretty big piece. And, I might add, relatively fresh.

All day long, people pitch stories to the media. It's like a big Hollywood cattle call, a casting couch that is always taking auditions. Somebody writes up a one-page press release. There's nothing special or official about it; anybody can write up a press release. ("Our church is having a bake sale for a little boy with cancer.") Will the story make the paper or the local television station? It will or it won't, depending upon whether the media consider the story important. Additional factors include whether the pitch is timely (". . . and the bake sale is next month! Don't forget to write about it and remember it for A WHOLE MONTH!"), whether it is a sympathetic hand that snatches the paper from the fax machine ("Oh, look at that poor little guy! He smiles just like my dear, sweet, spunky nephew. Darn it! I don't have any reporters since the legislature is in session. I know, I'll give that intern Melonie a chance. She couldn't possibly screw up a story *this* easy. Plus she's cute and it will give me an opportunity to talk with her.") or a cynical one ("Here we go again. What is this, a newspaper or Make-A-Wish? Geez, take a gander at this poor, ugly kid. He looks like Uncle Fester crossed with Gary Coleman. Well, tough luck. Churches can take out advertisements like everybody else . . . Whoah! Yowzah! Look at the new intern Melonie! Can't write worth a damn, but what an a--!! Hey, wait a second. I wonder if—? Hmmm. Eh, forget it. It won't get me anywhere."), or whether you can find a choice angle, like tying the story in with something larger. ("Timmy Barnes is one of the nineteen children mysteriously afflicted with cancer who attend Watson Elementary.")

Dumpster media manipulation is like any other kind of media manipulation. The point is that *the media is wide open to doing business with you, like a whore standing on the corner,* or, to phrase it better in terms of chaos theory, the media is *wide open to input.*

One individual, acting in a media-savvy manner, skillfully pitching information and angles he has unearthed, can single-handedly decide what the "issues of the day" shall be. I happen to be an expert in using state public disclosure laws (aka "the Freedom of Information Act," or FOIA), and besides getting a $30,000 out-of-court settlement from the Seattle Police Department in a case involving destroyed internal affairs documents,[3] I've unleashed just as many news stories using "open records laws" as I have unearthed from dumpsters. More, probably. Yes, more. But FOIA takes weeks at best, months most of the time, and often years. Half a decade is as long as I've pursued something under state open records laws, but I've heard of cases that dragged on for 20 years. Yet the Dumpster Freedom of Information Act of Diving—ratified and passed into law by the publication of this book and, therefore, with the blessings of the Dumpster Goddess—unleashes documents instantly. You can find, figure out your own angle, and fax in *forty minutes!* Instant activist gratification is such a rare and beautiful thing.

Yes, you can get inside information in all kinds of ways. You can get it because you work for an evil corporation ("Would you like to hear about some old chemicals buried near Watson Elementary in the 1950s? I saw it in some old files . . . ."), or you can use the FOIA, or you may have access to an informed source who just LOVES to chat with you on the phone about the married president she's been blowing, or you can . . . .

Go through their dumpsters. No particular method is automatically superior. I've come up short using FOIA and come up hot diving dumpsters, and what we're looking at *here* is dumpsters. Dumpsters certainly have a more random and interesting quality. I would compare it to a bungling burglary.

Let me draw out the illustration. Say, for example, you broke into the offices of MedEvil, Incorporated. You heard rumors of a training video documenting the details of a horribly botched third-trimester abortion, and you figured that tape was worth a *fortune* to the pro-life movement and could change the entire tone and substance of a national debate, maybe even change the outcome of a presidential election!

Your beliefs or politics are not the issue here and never were; not in this book and not in the last book. We are all evolving, and *extremes* are the exciting, leading edge of evolution. We might just as easily go to the other end of the political spectrum and say that you are breaking into the church offices of Telly Vangelist because you heard rumors of a videotape in which he fucks a thirteen-year-old Cuban prostitute in the ass, all the while telling her to repent for her sins (especially the sin of abortion), and you know that if Telly Vangelist goes down, it could affect the outcome of a presidential election!

Does the world turn upon so little? It does. A close and interested study of history reveals that time and time again, fate cornered on a dime. The battle of Gettysburg began over a foray for shoes. The prison riot at Attica started with a poorly welded iron bar. Bill Clinton was dragged through the mud over a stained dress. Personal history, like a chaotic fractal reflecting the patterns of the larger world, operates the same way. Who cares about the king who cried out "my kingdom for a horse"? My kingdom for ten more minutes on this phone card, for a gallon of gas, for her to look at me. *Little things mean a lot.* And if you're looking for so little, dumpsters are a good place to start.

So right about now, in this illustration, you're breaking and entering the medical or church offices. But then your flashlight battery dies. You've only got time to grab a knapsack full of documents before making a quick exit. So you grab in the darkness and stuff and stuff, but then you hear a damned siren and you are out the door, giving whole new meaning to the phrase "crash bar," and miraculously ("Thank you, Jeeeeezus!") ("Praise the Goddess!"), you escape.

Whew. That was close.

Dumping out the contents of your knapsack upon the living room floor, this is what was obtained from both offices:

- weekend cleaning check sheets (100)
- photocopies of newspaper clippings
- a mass-marketed videotape entitled "Florida's Number One Retirement Community"
- Original copies of angry, critical letters
- a broken porno tape entitled "Eleven Maids a Milking"
- and, finally, a torn photocopy, most of the second page of some kind of letter . . . .

. . . but the point was I didn't feel comfortable being in the room with him after the incident Friday. I wasn't trying to avoid doing my assigned tasks, I just didn't feel like getting fondled again when I turned my back. When I came back and found Mr. Ruby looking at files in my desk, I was so upset I asked, "What the hell are you doing?" To which he replied, with his usual 'Boy Genius' sneering smirk, "Going through your drawers, Miss Strauss."

At which point I . . . .

Arg! That's it! Where's the rest of the page? How did we find HALF a page? Maybe it got caught in the zipper.

The purpose of my "bumbling burglary" illustration is to show the odd, random, yet often fortuitous nature of dumpster diving for information. What we found, together, is exactly the kind of haul you can expect when you're going after dumpster dirt.

Let's pick through our catch, shall we?

Well, a bunch of cleaning forms. These will make great scratch paper. I've never known a dumpster diver to run short of scratch paper. Now, if these forms were really evidence of our *burglary*, we wouldn't keep them around, but since they're only evidence of dumpster diving . . . .

What else have we here? Newspaper clippings. Well, here's this organization we're so adamantly opposing, digging through their dumpster looking for something to fight it with, and we've never even read these articles. It might be worth our time to do so, wouldn't you say? In fact, anybody who uses *only* dumpster diving to fight these battles is shortchanging their cause.

As long as I'm permitting myself to go on tangents, I might point out how *frustrating* it is to deal with so-called advocacy organizations whose members don't even read the morning paper. To use a thinly disguised example, many's the time I've called Wary Ed Halden of the organization People Against Police Being Asshole Guys (PAPBAG).

"Ed!" I've shouted, into the phone. "Another young man shot dead by the pigs over a pack of smokes. FOR HOW LONG WILL THIS GO ON, WARY ED HALDEN?! Rumor has it somebody is organizing a march. I figured you might know who—"

"WHAT?!" Ed cries. "The police shot *another* guy? When did *this* happen, John?"

"This weekend, Ed," sigh I.

"Where did you hear THAT?!" Ed insists.

"Um, Ed," I say, trying to be tactful. "It was in all the papers and the television news programs and—"

"Oh!" says Ed. "Well, I was teaching at the Bible camp all this weekend and I'm still going through my stack of papers . . . ."

Needless to say, when the corporate and police cocksucking media swine want a spokesperson to comment about a people's march against a police shooting, they will quote Wary Ed Halden, who will—as always—appeal for people to remain calm "until we know all the facts."

This tangent springing from the discovery of a hypothetical newspaper clipping within an imaginary example is just my sweet way of saying . . . .

Be wise. Be informed. Know and study the enemy so that when you obtain a piece of dirt in the dumpster—or by any of the many other effective sources—you will be able to maximize its impact. The smallest piece of info-crap has the potential to be brandished like a broken bottle in a street fight. And a whole shopping cart full of highly classified documents is nothing more than a heap fit for the recycle bin if the diver is unable to recognize its value and potential.

Where were we? Hmmm, some kind of videotape about a retirement community. Sweet! We'll just pop it into the VCR for ambiance while we sort. How many pieces of *other* people's lives make up the life of a dumpster diver! Aaaah. Soothing music. A golf course. Old men walking around in shorts.

This pile of angry, critical letters looks interesting. These people are all potential allies we can, in a careful and subtle manner, bring into our struggle. Some of these folks might even have damaging information. As for the broken porno tape, its presence in the office and/or the dumpster is interesting. But it is a red herring. Maybe it belonged to the recently fired janitor in the medical offices. Maybe Telly Vangelist had an aide doing research about the evil, fornicating world out there. If we *really* found the tape in a dumpster (instead of in our hypothetical Watergate homage), perhaps somebody was cleaning out their van in the alley and threw it into the trash.

Stuff ends up in dumpsters in all kinds of ways. It is a rookie mistake to *assume* that because something is in a dumpster it was the business or person who actually *uses* that dumpster who put it there. Flexibility and adaptability, the ability to think on your feet, is the key to success in this quasi-biz. As with everything else.

But I'll save that for the "fortune cookie chapter."

We've been saving dessert for last. The torn letter. We peer and peer at this piece, trying to infer, guess, extrapolate, theorize—all the things the human mind does to put broken pieces together.

Be flexible. As a character in my still

unpublished novel says, "I believe all things, I doubt all things, I constantly weigh my doubt against my belief. Some things I believe 99 percent, like the fact my parents are really my parents. But if I were to learn otherwise I won't go insane, because I leave myself an intellectual escape hatch."

What if this torn page is part of a novel somebody is whipping together upon their word processor? What if this is an old sexual harassment matter which has *already* surfaced in the media? A piece of something hot, sure, but five years ago.

But let's assume this is a relatively fresh episode and we are holding a piece of it. Let's further assume that we can figure out who the two participants are by paging through the telephone book and a corporate/church directory. What if Miss Strauss already has an attorney and they are preparing to file a lawsuit? If we make this thing go public in a premature manner, it could fuck up everything and actually *hurt* our cause. Hmmmmm.

As dumpster divers, it is often our nature to *dive right in* and *get our hands dirty*. But we need to be aware of this tendency, which is usually a strength but can sometimes be a weakness. Unlike, for example, a television that might be obviously broken or obviously working, the value of information obtained in a dumpster can be much more difficult to discern. While you contemplate the fragment in your hand, wondering if Miss Strauss might have inside information about the tape you were actually after in this little foray, and whether you might find a good way to contact her ("Hello, babe, I'm a guy who goes through the trash, and I hear you let dirty old men into your drawers."), the Florida tape continues to provide ambiance . . . .

"You've worked and saved all your life, making good investment decisions, always staying on top of the game. How would you like to make one decision that can keep you on top of the game for ten, twenty, or thirty golden years?"

Reporters love to gossip. I've had reporters pump me for information about, for example, police internal affairs scandals. Their interest was so *intense* that I half-imagined I would see something in the next morning's paper. I've seen scandals break and received e-mail from those same reporters. The message was always basically, "Have you heard anything?" I figured an identical message was going to dozens of the reporter's regular sources of gossip. Feeding pieces of gossip to reporters may not produce immediate publicity payoffs, but media manipulation is like a blackjack table where you play with free chips, chips that you can cash in for real value. The only risk is to a person's credibility if he/she just gets too damned excited about a piece of nothing.

Here's a recap of a conversation that I actually had with a reporter from a little paper in Washington state. I had never met this reporter before and, indeed, I only knew of his publication from flipping through the phonebook, looking to see if the little town in question had a local paper.

"Hi, my name's John Hoffman. I don't know if you've ever heard of me, but I'm kind of well known in some circles as an author and activist, and recently I dug up some information I thought might be of interest to you."

"Oh yeah? What do you have?"

"Well, you know the technical college in your town?"

"Yeah, [name of college]."

"Exactly. Well, I was rummaging around in a dumpster . . . that's what I do, you see. I'm the author of *The Art and Science of Dumpster Diving* and I was out with a reporter from the *Seattle Times* . . . ."

"No kidding? You wrote that book? I think I've heard of that book, actually. Yeah, yeah, my cousin! He's a big dumpster diver and he's got a copy of your book!"

"Really? It's a small world after all. Well, hell, let me tell you what I turned up, then, since you know I'm good at this kind of thing . . . ."

"OK!" (Now I can practically hear him salivating)

"It seems that your technical college is using some kind of consultant who has an office in my neighborhood, and they discarded this document. It's a sort of an audit of the automotive sciences section. Well, according to this audit, the technical college is running a serious risk of violating a state law because they've got [name of BIG corporation] in there letting the students work on the vehicles, calling it 'education' when it's just supervised work."

"And they might be breaking a law?"

"Yeah, it's actually [here I cite the law, chapter and verse], and it's right here in this report. There's more, too, like some lax accounting procedures and—"

"WHERE ARE YOU? DO YOU WANT ME TO COME AND GET THIS?"

"No. Actually, it's only a few pages and it's only one stamp, and I've got it in an envelope right now. I just wanted to confirm your address. Besides, I don't know how important this is. I mean, I look at it, and I think, well, this seems sort of scandalous. But for all I know, your paper might have already done a story . . . ."

"No, first I've heard of it!"

Etcetera, etcetera.

Please note how I established credibility. God, there is nothing more annoying to a reporter than citizens who think they're Deep Throat just because they've got a few secrets about their recent former employer or a desire to talk "off the record" when discussing relatively minor bullshit.

Cut right to the headline. That audit was full of tedious details, but the *hottest* matter was a violation of state law. This is part of "pitching" the story. You need ONE BIG ANGLE you can sum up quickly, in a sentence or two. Enough to convince the reporter to look at everything else.

I leave it to others to decry "sensationalism" and "lack of content and context." News is just as much information as entertainment. How often do I really need to read news stories to make important decisions? I probably won't be traveling to Cuba anytime soon, but I'm still interested in what is happening there. I might say I am "informing" myself, but really it's just a very cerebral form of entertainment. So if you want to get a story publicized, YOU GOTTA BE INTERESTING! YOU GOTTA BE ENTERTAINING! YOU GOTTA UNDERSTAND SHOW BIZ, BABY!!

If I had one piece of advice for activists who want to be effective, it would be this: QUIT BEING SO BORING AND SELF-RIGHTEOUS! The reason I can single-handedly terrorize banks with nothing more than a knapsack full of documents is because I'm having so much FUN doing it. The media know that I'm their little Nielsen Week Slut Boy. I'll eat discarded movie popcorn right out of a big black bag when that camera is live, live, live, and I'll hold still while they zoom for a tight shot on my ass while I'm going down on a dumpster. If you are full of positive sun-shiny love energy while you explain to the media what you've got to share, they will be unable to resist your animated face. They will love you even if they don't run with your story every time you pitch one. How do you think Julia Butterfly Hill got so big? [5]

Conserve credibility. Be aware of what you know for sure and what you don't.

You *know* you have a document you found in the dumpster. (I try to avoid being vague about where I obtained the document, but it might be unavoidable.) This you know for sure, and everything else is speculation. I recall being caught up in the details of a fascinating lawsuit I found near the University of Washington until I saw it had been filed in "moot court." (For those unfamiliar with the term, it was a *practice* lawsuit used in the schooling of lawyers.) What you have might be interesting, but it might not rise to the level of a news story. It might be newsworthy but no longer timely. Some information, like the dairy products you learned to salvage in my first book, has a short shelf life. There's a saying in journalism that may be the wind at your back or the fart in your face: *old shit doesn't stink unless you stir it up.*

And, of course, there is always that *incomplete* aspect. What if the missing pages of the letter go on to explain how Miss Strauss and Mr. Ruby worked out their differences and established a respectful, even affectionate relationship?

But these problems are not limited to dumpster diving. Anybody who tries to dig up information by any means deals with problems of value, timeliness, completeness, and a host of other factors. As well as any other kind of visionary, a dumpster diver can—by pure will and a spark of brilliance—determine what the "issue" of the day" shall be.

I know because I have done it many times. But I have done it both through my ability to scavenge *and* my rather advanced ability to engage in media manipulation.

"What could be more enjoyable than taking your meals at the members-only clubhouse after a nice round of–"

Bssssssshhhkkkk!

"Smile for the camera, sweetheart. That's not a very big smile, dear. Is this thing working? OK, the date is May 1st, 2000, and it's about three in the afternoon and we're about to go in, ha ha ha–"

Your jaw drops. Somebody has recorded something in the *middle* of this tape. And it looks like– *It looks like–!* IT'S A DISCARDED COPY OF THE VERY VIDEOTAPE YOU WERE LOOKING FOR!

You can change the whole world. All you need is a little luck, and this is a game you can play for free, baby. You are holding the butterfly of chaos in your hands, and you can feel the flutter of wings that will unleash hurricanes.

• • • • •

The other side of the coin. An overwhelming volume of information.

I once dived the headquarters of a major insurance company, a secured dumpster inside a parking garage. The funny thing was, all I wanted from that foray were a few acid-free file boxes that I could see in plain sight. But I soon realized I was looking at, well, summaries with attached documentation of probably every insurance claim that was filed with that company from approximately 1971 through 1985.

I wondered if they had put all this stuff on microfiche or CD ROM or something. So much information! Literally tons of paper in a super-sized dumpster more suitable for a construction site (and, I might add, it wasn't a designated recycling bin either).

In only a hundred years, some of this stuff might be *fascinating* to the children of the people named in these forms. How often have I stared at a yellowing land deed or a citizenship application preserved on microfiche, seeking some fascinating tidbit about my ancestors. How seldom does our society think, "What about in ten years? A hundred? A thousand?" before throwing something in the landfill to satisfy the desires of *this red-hot minute,* whether it is an old hospital record or a fluorescent light bulb.

If you dive dumpsters looking for dirt to use against the powerful, something you might encounter is the OVERWHELMING WINDFALL OF VOLUMINOUS INFORMATION. And then what the hell are you going to do?

Here's what. Get yourself a sponsor who will help you obtain a scanner to put that stuff on CD-ROM. Get yourself volunteers who will scan the documents, one by one, so you won't be paying a professional service 12 cents a page or doing all the work yourself. And then get yourself set up with a website and find ways to keep your ass legally covered. There are people who have done this very thing, including–why is it these examples *always* seem to pop up?–a guy who obtained approximately 10,000 records from an abortion clinic and put them on a website.

I myself was involved in this nation's largest release, via CD-ROM, of 6,000 pages of police internal affairs documents, but that involved five years of FOIA requests and not dumpster diving.

Well, not *much* dumpster diving. There was, of course, a special file I call "Roswell" because it fell out of the heavens into my eager hands and sent me flying off in directions I never thought of going, but *those* details have yet to be declassified. I charged $200 for the CD-ROMs and mostly sold them to lawyers and the media. So, if you find an amazing windfall of data, it might be possible to make a small fortune by selling it in one form or another or at least raising a lot of hell by making it readily accessible to the general public.

Sure, some places have shredders. In fact, there was a Wisconsin company involved in "professional document destruction" that wanted to hire me as a consultant! They never tendered an actual cash offer, so god knows what my answer might have been. It could have been a lot of fun, jetting all over the country, speaking to gatherings of corporate executives.

"I am your worst nightmare. I'm the little guy your company just screwed over, and I just read a book called The Art and Science of Dumpster Diving . . . ."

Maybe, along the way, I might have urged them to do a little more recycling, a little more donating to food banks, and a lot more pay for their lowest paid workers. My point is that shredders, even where they are present, don't

John Hoffman poses with an anonymous donor who kicked in almost $2,000 in order to scan to CD-ROM some 6,000 police internal affairs documents, accumulated as part of a five-year effort by Hoffman against the Seattle Police Department. The release of the documents tore open numerous scandals, documented mostly in the pages of *The Seattle Weekly* by reporter Rick Anderson. The easily reproduced format of the documents continues to have an impact on the department and serves as a model of how to reform the police by systematically dragging their scandal documentation into daylight over a prolonged period of time.

always get used, or they have a week of downtime. Some documents, incredibly, get through overloaded shredders partially intact. Sometimes people don't have time to use the shredder, so they throw away bags of documents designated to be shredded. In fact, some places rely upon padlocked dumpsters instead of shredders. Tee hee!

If you find a few fragments, manipulate. If you find a windfall, computerize! Or haul it off in a truck and, at your leisure, look for something hot. But I continue to say that it is a shame so much documentation, that involves so much human effort, which might contain so much data hidden *between* the data, gets tossed in the landfill or made into newspaper pulp, while we go to museums and peer, fascinated, at fragments of Babylonian wheat records behind bulletproof glass. In a mountain of old speeding tickets we might find the identity of a serial killer. In the canyons between mountains of credit card receipts, we might find red flags that herald the outbreak of a plague.

Fans from all over the world have sent me newspaper clippings in which documents discovered in dumpsters blossomed, beautifully, into major scandals. Often, the fact that documents or something else has been discarded *is* the scandal. I have to shout out, once again, to Craig S., who sent me one of my favorite articles of this nature. In that particular instance police had thrown away old files that included, for example, photos of murder victims. (Speaking from considerable firsthand experience, if you go to the media with this kind of story, be prepared to see "super" security at that particular dumpster in the near future.)

The rise of the Internet creates opportunities not just for divers with huge windfalls of information but those with mere fragments. We live in a world where major corporations have their own "sucks dot com" websites. I know that a lot of divers out there don't have computers quite yet. I would advise you to go get yourself one or you will turn into one of those life-forms that can't keep up with the curve of evolution. And, at the risk of writing in what seems like a foreign language, you can scan important documents you find and, using a hotmail account, send them as attachments to a "sucks dot com" or similar website. I think print media is still preferable, but hot Internet chatter can actually attract the attention of the print media and set off the very same publicity firestorm you tried but *couldn't* unleash with just a few dumpster-dived documents.

(Particularly, I might add, when the reason you can't get your story into print isn't because you're bad at playing the game but because the media are playing *their* little game again: acting like advocates of the public interest while sucking police and corporate cock.)

If you're doing this kind of thing openly, you might get a call from a reporter wanting to know if you're the guy who unleashed the infamous "Bloody May Day" video upon the world, the same one playing this very *moment* on C-SPAN.

In my previous book, I told people NOT to talk to the media about dumpster diving, saying that it only unleashes anti-dumpster-diving programs. To some extent, this is still true.

However, the wave of "dumpster consciousness" unleashed in the early 1990s has brought this activity way the hell out in the open. I believe that dumpster consciousness has helped our planet move toward a future utopia: self-contained cities miles high, built upon arches that straddle grassland and rainforest where cities once stood, filled with nearly angelic beings who live upon light and send e-mails to planets around Polaris. Nothing wasted nor cast into the pit, not food nor plastic nor electricity.

That is what we are moving toward when we salvage a perfectly good crate of oranges, when we preserve discarded papery data and send its essence around the world upon beams of light energy. We mere dumpster divers are part of this exciting evolution toward the stars and the angels.

On the downside, however, there are some cretins who hate dumpster divers, and publicity about this activity catches their attention. Self-serving city attorneys looking for scapegoats so they can pass new laws to shamelessly pander to middle-class voters. Bored, sadistic security guards. People who never got past the trauma of shitting their pants in second grade. Consultants marketing pieces of their own paranoia. Media sluts who have to suggest a "course of action" to justify their own sensationalism, when they were perfectly happy to be allies with the dumpster diver hours before . . . .

Never will I forget the Spirit of Seattle during the uprising against the World Trade Organization. The police rounded up 150 demonstrators and put them on three busses . . . and the demonstrators took over the busses and wouldn't get off. As the hours passed, as the night grew cold and rainy, lawyers tried to bring the prisoners bread and water, but the police wouldn't let the rations through. So the bread and water sat in the pouring rain, and those brave, handcuffed children waved and shouted.

"We love you! We *love* you!"

I think of your bold, beautiful faces . . . the mysteries you will solve, the evils you will expose to the purification of light. But when my little boy—Alexander James Hoffman—falls down and skins his knee, I wish it could be *daddy's* knee injured instead of his. Likewise, when I hear that my mother has been ill, no matter how minor or brief the illness, the realization that my mother is mortal—that one day, sooner than I realize, I may be *motherless*—these things make my heart skip a beat.

The thought that you might be captured by the enemy, interrogated, charged with some ridiculous crime—it absolutely makes me sick inside. How well I recall when a newspaper in New Jersey called me for commentary about a diver who had been arrested for going into the dumpster of a cosmetics company, making himself and his family decent money on the side by selling those ridiculously overpriced cosmetics at flea markets. The newspaper didn't quote the part where I mentioned the fact how that company tortures animals to test cosmetics, therefore they deserve whatever they get. The paper did, however, quote the part where I said we have a serious solid waste problem in this country, and those divers "deserve a medal."

Engaging in a bit of profitable media manipulation, I sent the clipping from the paper in New Jersey to *The National Examiner* and made $100 promoting my book and opinion . . . .

But *please* be careful.

How well I recall the sight of my little son, only 48 hours old, wedging himself in the corner of an incubator, fighting off three nurses who wanted to hold him down and stick him with a needle, flailing one of them so hard with a trailing length of heplock tubing attached to his little wrist that she stuck herself with the needle intended for him!

Good god, I thought, isn't there a more *evolved* and *humane* way to perform this test than sticking a needle in a baby?

"GIVE 'EM HELL, SON!" I shouted, hardly realizing I had spoken, let alone *shouted.*

"You," said one of the nurses. "Out!"

The advice in my previous book is still good advice. The media sword can cut for you or against you. If you don't want your dumpster to dry up, then you should conceal what you are doing as much as possible.

(Before the end of that conversation with the reporter in Washington state, I asked him to keep quiet about the origin of the documents. And this sort of reasoning seemed to work really well: I'm not really *connected* to this story. I've never even visited this college. The documents just fell into my hands, sort of by accident. I'm not trying to be Deep Throat or something, but if you'd make an effort to keep me and my dumpster diving out of this story, I'd really appreciate that. And who knows? I might find something else for you in the future in that same dumpster.")

But it is important—*it is vital to the survival of the human race and the life-forms on this planet*—that corporate and government evil be exposed. That is why I'm telling you how to deal with the media in a wise, effective manner.

But please don't get hurt. Because I care about you.

Yet here are new marching orders, if you will have them.

GIVE THEM HELL!!!!!!

CHAPTER 4

Dumpster Diving Hollywood

We talked, in the previous book, about you becoming a movie star, didn't we, babe?

Even if you loathe that sort of thing, don't *think* about skipping this chapter. I'm going to show you how to make a "Hollywood" buck or two, even if you live in a town whose main industry is poultry processing.

We'll start with warmed-up leftovers from the last chapter, a sort of bag lunch on the bus to La La Land. This is how you turn a book about dumpster diving into a worldwide media event. (The point of this stroll down the Hollywood Walk of Fame is not to relive past glories, though god knows that's always fun, but to show you where certain valuable items you might have overlooked can be hiding in your town, as are new ways to market those items that have emerged in the past few years.)

First, using a contact or two, you land a little blurb (i.e., short mention) in a limited-circulation satire magazine in San Francisco called *The Nose*. Then you send that article to a whole bunch of *other* media, along with the usual press releases. These other media say to themselves, "Hmmm, looks interesting, and this little edge culture magazine seems to have picked up on it early, as they often do." So a few *other* publications write articles too. Some of these publications have wire services, which magically zap the article to papers all over the country. These local papers want local commentary, so they seek out a local diver and then want a quote from the "expert" author. This process continues, infinitely, until finally the dumpster guru is doing a weekly radio show in South Africa and having rand wired directly to his bank account, and I am NOT making this up.

But all along the way, you gotta keep fanning the flames, creating a feedback loop. If you've got five articles about something you're doing, you should be able to land a television interview with little difficulty. God knows you don't sit around waiting for things to happen; you *make* them happen, prying open the lid of dumpster destiny and claiming what is rightfully yours . . . in this case, a fortune to be made off certain discarded items you might have overlooked.

"Oh my god, one of our English majors wrote a book!" This seems to be what Concordia College in Moorhead, Minnesota, was thinking when they scrambled to arrange a television interview for me with a local station. *We've got a successful English major on our hands. Reeeeeeeeeeenk! BATTLE STATIONS! BATTLE STATIONS! BETTER PUBLICIZE IT SO WE CAN KEEP RECRUITMENT LEVELS UP! GOD KNOWS IT'S HARD TO SELL THOSE LIBERAL ARTS DEGREES!*

Me and my beautiful wife Tina traveled from our home in El Paso to an NBC affiliate in Albuquerque, New Mexico, where we were interviewed, via satellite, by a newscaster in—good god!—Fargo, North Dakota.

"How much does satellite time cost?" I asked one of the techies at the station, and he named a figure that was *bigger than the advance on my book.* While it's no revelation to realize there's a racket everywhere you turn—somebody is always making money, somebody is always losing money—be advised that the chaos you create is soon converted into somebody's racket. What, for example, was happening in the mind of some assistant corporate cog inside that bank I humiliated during Nielsen Week?

Smedley, the bank manager, has been disgraced by the dumpster document fiasco. If I can only sidle up to CEO Fenster in the men's room and get him to say something to me, anything, then I can get him interested in all the dirt I've been filing away about Smedley and make my move . . . calm, eyes on the wall, make some "relief" sounds, clear the throat . . . .

"Did you get a copy of the interview?" my editor at Loompanics demanded, breathlessly.

"Hell yes I've got one," I said. "I couldn't get anybody at the station to tape the interview, but my mother-in-law taped it off the television. It's in the mail on the way here."

Off went a dozen copies of that awkward first interview on a rinky-dink Midwestern station to "info-tainment" shows in Hollywood. All the time they are looking for some angle, some new controversy, some exciting video. Their appetite is insatiable. You just gotta keep pitching yourself. Sure enough, the bait worked and we got a bite: "The John and Leeza Show Live from Hollywood."

Later, the show would become half of itself standing alone, "The Leeza Show." Three MORE trips to Hollywood would follow, including an appearance on "To Tell the Truth" where I won $1,000 for a few hours work and—newly single—hit upon the gorgeous blonde debutante who paid $4,000 for a date with Gary Coleman.

Nice work if you can get it.

Yes, now I've dived Hollywood. I've dived the "Star Trek Deep Space Nine" dumpster, for crying out loud, I've dived talk show host Arsenio Hall's dumpster, I've got Polaroids of Pee Wee Herman (NOT making this up), and for a long time, I had glossies of would-be stars and starlets with resumés attached on the back, including one of those unfortunate "Star Trek" guys who wears a red shirt and never gets more than a few lines before dying in the opening scene.

If you have dreams of becoming a movie star—and hey, there are worse ways to make a buck—let me tell you, it is a revelation to see the actual resumés (attached to glossy black photos) of people trying to break into the big time. The shampoo commercials, the bit parts in state government films about hepatitis, the summer stock theater roles, and, almost always, some monetary investment in a school of acting

Hollywood isn't so hard to reach.

Buses go there every day.

It used to be really difficult to sell shit that you found in the dumpster. I guess I didn't know how difficult it was at the time because, hell, I was making money here and there. When I came back from Hollywood with a "Star Trek Deep Space Nine" script as a souvenir of my dumpster diving on the Paramount lot (all under official escort, I might add), it seemed natural to raffle it off to support the local science fiction club. I mean, how many diehard "Star Trek" fans have any kind of serious money? And how much of my life do I want to dedicate to seeking one out who may be the exception to the rule just to turn $25 into, say, $50 or $100? Raffling off the script was good, clean fun, and so was doing a lecture at the convention called "I Dumpster Dived the Star Trek Dumpster."

Nobody will come up with a "Star Trek" angle lower than THAT. It makes me perversely proud, in a way. To think that I found a "Star Trek" photo novel during my first solo dive behind a bookstore, and here I was diving the actual "Star Trek" dumpster in Hollywood! The tadpole has evolved into a prince . . . .

I enjoyed that moment, the fun, the campiness of it all . . . .

But *today* I'd sell that script on eBay.

There's no getting around the fact some of my youngest fans were way the hell ahead of their so-called guru with *this* particular thing. Let the historical record reflect that I first read about selling dumpster-dived stuff through eBay in a zine called *This Month* by Jodee in Brooklyn. At the time Jodee was mailing me copies of *This Month*, she seemed no older than 18 or 19. Unlike other dumpster fans who are eager to establish prolonged contact, Jodee seemed content just to mail me her zine, which periodically mentioned scavenging various items, then selling those items through eBay.

"What is eBay?" I thought.

Since I have some experience in the military, I pictured the "bay" of my military barracks in San Antonio, Texas, and young men in formation snapping to attention and calling out the alphabetic designation of their bay, along with the offensive sexist slogan of the day.

"C Baaaaaaaay! Smells like fish! Tastes like chicken!"

(Then all of us "ate-up maggots" from C Bay would have to drop and do push-ups in the sun for a very long time, while half a dozen pay phones rang and rang in the hot sun because some soldier had cleverly used a cell phone to send out dozens of pages, all with the numbers of the pay phones, while the drill sergeants were just, like, clueless. God, technology is FUN! If I weren't a dumpster diver, I would want to be a hacker or a phone phreak or some fun new thing being invented right this moment . . . a fluxus techno pranker . . . .)

ANYWHO!!!!! The word eBay made me picture a military barracks, open to the sunshine, where people like Jodee would gather to sell their wares. A kind of flea market, neat and orderly. But when I have a question about something, it tickles at the base of my brain until finally the answer makes its way to me. Within days of reading the phrase in Jodee's zine, I read what might have been the first of many "eBay controversy" wire stories. Today, these stories are a mainstay of the morning paper. Now enrolled in college again, at the University of North Dakota, I actually attend a philosophical discussion group called Electric Mind with a young man who sold his soul on eBay.

Human kidneys . . . artwork by mass murderers . . . models selling their ovum . . . and, yes, dumpster goodies galore!

How many times have I found some odd little item—for example, the secret bylaws of a fraternity carelessly discarded after a night of failing to recycle LOTS of aluminum beer cans—and I thought, hell, I know this is worth a lot to *somebody*, but how do I *sell* the damned thing? You just *know* there's a buyer out there willing to pay big money for some rare and obscure item in question, but the *effort* required to find the equally rare and obscure buyer seems to cancel out the value that might, ultimately, be gained.

But no more, friends and neighbors, children of all ages! I tell you that eBay and services like eBay are the answer to your problem! Why, eBay can make you a fortune! We must no longer be mere divers; we must be *cyberdivers!*

Those of you who have been divers for a while know that odd collectible items—funky clothing, old books or albums, the stuff which is the staple of profitable yard sales, the same yard sales you can have after collecting enough junk from dumpsters—are *also* a staple of eBay commerce. I would urge you to move in this direction. If you aren't "wired," or the only computer experts in your home are under the age of twelve, then have your twelve-year-old assist you, or ask a wired friend to help you sell a few items on eBay. Once you realize the size of the market you are reaching—their relative affluence and willingness to spend money—and how that translates into MORE MONEY, you will never turn back. (Of course, you need a credit card to have an eBay account, which, for some of you, will suck.)

The very consignment stores, antique dealers, and flea markets where I urged you to bring stuff back in the mid-90s are now getting wired themselves. If *you* aren't selling your dumpster-dived goodies this way, odds are somebody else is selling them *for* you and making a profit.

There is also something wonderful about all those discarded items ending up in the hands of people who want them *so badly* that they would go to such lengths searching for the item with a beam of angel light upon a worldwide network of computers. After strange and twisted travels . . . after being the chattel of mysterious strangers with menacing accents . . . after being cast aside and left for worthless, then miraculously raised from the depths . . . at long last the cookie jar, the red bike, the little figurine has found its way back to hands that tremble with love and desire.

It's a beautiful thing.

But, of all the crap sold on eBay (oops, I mean beautiful things finding their way to hands that tremble with, etc.), the stuff with the highest margin of pure, lucrative, almost *magic* profit has got to be "personal memorabilia." Let's call all of this stuff "Hollywood" even though we might be talking about, for example, a letter—worth quite a few bucks—from a teacher who is famous for having sex with one of her students, and not a Hollywood movie star per se.

Scripts . . . correspondence . . . resumés with attached glossies like we talked about before. Hey, if it belongs to a sports figure, a famous criminal, a movie star, a retired movie star, a porno star turned movie star turned retired movie star, if it would be nothing but crap good for a moment of snooping and nothing more *except* it belonged to somebody famous and so it's actually worth *money*, but god knows it would be hard to sell without the miracle of eBay, eBay, eBay (or better, any of the very specialized electronic auction sites that deal with, for example, television and movie memorabilia), then it's all "Hollywood."

Where is Hollywood? Let's start with where we know it is for certain, as our bus pulls into Los Angeles.

The security at studios like Paramount is tough. You might as well think of dumpster diving the Pentagon. Not impossible, but certainly not probable. If you work at a Hollywood studio you might have a shot at a dumpster, but why would you risk your coveted job *in* Hollywood just to dive a dumpster there? If it is possible to track the refuse to a dumpsite or solid waste processing facility and score *that* way . . . if the garbagemen pull over and try to salvage a few scripts . . . well, it probably happens, but there's an easier way accessible to almost everybody.

The movie industry is all over the place. The biggest cache of scripts I found wasn't at the Paramount lot but in a residential neighborhood near Burbank. Movie stars don't live at the studios. Some of them live in Colorado. Sometimes movies are filmed in places like South Dakota. All the scripts and correspondence and photos and bric-a-brac associated with making movies generates garbage, and this potentially lucrative garbage can be found virtually anywhere.

But of course the epicenter is Hollywood itself. So when in Los Angeles, I would suggest that you dive logical targets *peripherally associated* with the movie industry.

Agencies, for example. Or the garbage cans of people associated with movie production who might drag home paperwork. If you're going after the trash of a star you will probably run into security, but if you stay away from the hot spots (David Letterman's house, in particular. Madonna, Madonna, Madonna) and dive the cool, untroubled waters (a place that copies and binds printed matter located right in the heart of Hollywood), you might get lucky, turn up some scripts, and avoid a nasty encounter with private security.

As for other Hollywood goodies that might be

out there, the key is knowing your corner of the world and making brilliant, intuitive leaps.

You can also just get unexpectedly lucky.

Does the famous mass murderer write regularly to his mother, who lives in your town? (Mass murderers may be a bad example. You can bet Bundy's mother was conscious of her garbage since little Teddy mentioned how he got started on the road to evil by scavenging lurid "true crime" magazines from his neighbor's discards. *Hey kids, stay off the Road of Evil while dumpster diving.*)

When you read the newspaper, imagine it says TRENDS IN TRASH across the masthead instead of say, *The Boondocks Herald-Press.* As you read, think to yourself, "Hmmmm. The famous child star came to visit his sister. I didn't know she lived here in Crusty Fork, North Dakota!")

Anybody can walk up to a famous person and ask for their autograph on a dry cleaning bill. Autographs aren't worth much until the famous person dies. (STAY OFF THE ROAD OF EVIL, MY SON!!! STAY THE HELL OFF!!!!) But if you find a bona fide piece of that person's life—an old contract or wedding invitation, for example—it can be valuable and highly salable even while that person is alive.

Also, don't forget, *the word of a dumpster diver is worth its weight in gold-plated Oscars.* In other words, your *written testimony* can not only verify the authenticity of the item but add value. When we raffled off the "Star Trek" script, I wrote upon the inside cover that it was "lawfully recovered while under official escort at Paramount Studios." It would be better to write out your testimony on another sheet of paper, but why can't YOU be a star, too?

A nicely typed testimony, preferably on bond paper, stating that you did "lawfully recover" the item in question at such and such date and time can mean the difference between a valueless broken hockey stick versus a valuable piece of sports memorabilia.

(And do not skate upon the Thin Ice of Evil, my son.)

One thing I would suggest, however, is to unload the stuff while it is hot. I found the Pee Wee Herman Polaroids after the scandal had cooled, and I am *still* waiting for that talented, funny comedian to become a big star again so I can unload his pictures for a decent price.

So, what exactly is *possible* in this highly evolved, potentially lucrative niche of cybernetic Hollywood dumpster diving? Could you, for example, catch a bus to Los Angeles, spend a weekend dumpster diving for Hollywood, go home, sell the stuff through eBay or a similar venue . . . *and* make enough dough to pay for a semester of college?

This was simply not possible in 1993. Likewise, the ability to dive damaging information and use it effectively has drastically changed. The potential value of salable items as well as the power of damaging information has increased exponentially though the growth of the Internet. This new medium was *made* for dumpster divers.

And you were made to be a star, babe.

But why Hollywood, California? Why shouldn't we grasp new technologies to make our own movies—salvaging our props, our costumes, everything we need, finding amazing stuff and then writing a script *around what we have available?* Why not Seattle or Minneapolis or Boston, or some little town in the middle of a Kansas wheat field as the center of a new, brilliant, and economical movie industry? Or, to express it in Modern Geek: Hollywood is like a mainframe; why not every little town a PC?

The farm where I grew up in Minnesota was the most wonderful place in the world. We had tons of stuff from dumpsters to play wild, imaginative games. Anybody who has watched the antics of Jim Carrey has seen the product of a dumpster-diving family. And all the dumpster divers I've met are inevitably the same way: Super animated. Full of life. Almost living cartoons.

Have you ever said to yourself, "Damn, that routine my brother just did was as funny as anything in the movie theater!" Have you ever looked at a friend, relative, or co-worker and thought, "He/she is as beautiful as a movie star!"

Then to hell with Hollywood.

I make no claims of dumpster-based comic genius. I am, however, infamous for busting up my co-workers in fits of laughter with imitations based upon events in the workplace. Still, I'm not nearly as funny as my little brother. Which brings me back to my point . . . .

Why should the correspondence of a Hollywood starlet be more precious than, say, a letter from my baby brother? WHAT SCREAMING BLACK VOID ARE WE FILLING THAT WE MUST PAY THESE ENTERTAINERS MILLIONS AND MILLIONS OF DOLLARS AND EVEN THEIR CANCELLED CHECKS ARE WORTH MORE THAN OUR PAYCHECKS?

This I will say about Hollywood: digging through their trash certainly pops the bubble of glitz, and that's all for the better. As for criminals and sports figures (to take a few more shots), I think that everybody is capable of great passion and fantasy, not just those who act upon it by, for example, systematically strangling hookers. I think a man who does a belly flop into freezing water to save somebody in a sinking car deserves more praise, more applause, and certainly more *money* than a graceful athlete who throws a ball up in the air over and over like an *autistic child.*

Make a buck if you can, babe, a buck to seek higher education or put decent food in front of your children by finding Hollywood crap and selling it to the highest bidder. If you find something that pleases you and you want to keep it for yourself, take pleasure in that object and the clever way you obtained it. If you have a chance to go to Hollywood, to meet stars and be on television, enjoy that as much as you can because you'll find that other people will be more amazed with you than you will be with yourself. You will feel like you've seen Daddy putting the presents under the Christmas tree in a nicotine-stained undershirt while all the other children say, "YOU SAW SANTA CLAUS?! THE *REAL* SANTA?!"

Hollywood is a place where you eat Erie cheese and wash it down with SlimFast in the "green room." When the show is done, all that good food—black olives and chicken wings and little gouda cheeses—all gets thrown away while the info-tainment news shows blather on about the plight of hungry Americans. Hollywood is a place where the coat of the talk show host has no lining so he can wear it under the hot lights—a "coat costume." On television and in your video store, mainstream media and hard-core pornography *seem* to be two different and distinct worlds. In Hollywood, it is amazing how those two worlds mingle freely. You eat Chinese food and you're hungry an hour later, allegedly. I've never found THAT to be true. But THIS is true: You eat food in Hollywood, you stuff it in your face, not knowing where your next meal is coming from, and yet you are STARVING the moment you finish it, and a bit faint and queasy at the same time, just at the THOUGHT of what you put in your mouth and swallowed.

That is Hollywood. The garbage and the dumpsters are exactly the same, but some of it is worth a pretty penny.

CHAPTER 5

Dumpster Fan Feedback

We covered a lot of great diving spots in my first book. If you want to find out how to salvage perfectly good produce from grocery stores; if you want reflections about Thy Neighbor's Dumpster, and how all of the goods and services of our society are eventually found in people's homes and eventually in people's dumpsters; in fact, if you want details about the best, most common, and most convenient places to dive, *TAandSoDD* is still the authoritative text for you to read. One reason it took me seven years to write a sequel is because, damn, at that point we had things *covered.*

But, thank god, people read books critically. And one piece of feedback I received over and over was, "What about rental storage facilities? I dive there all the time and they're hot!"

I thought I had covered those places when I mentioned how, when people move, they throw away a lot of really great stuff. In retrospect, I realize I should have gone out of my way to mention not only apartments around the time the calendar page is about to flip over, not only the college campus in the spring, but these rental storage facilities.

Space. We wall ourselves into spaces and charge a fortune for the use of these spaces. The magic box shrieks its constant demand that we buy, buy, buy more goodies, and then we have to cram it all into our limited living spaces. Somebody builds a new facility and suddenly the *spaces* around that facility become more costly, and people have to move—making decisions about what to keep, what to throw away, and, as time and space collide, their decisions grow increasingly frantic. *Well, this headboard is probably an antique, but it's not a family heirloom, and even though I paid money for it I never liked it that much . . . .*

Like oversized animal cages meant to hide, not display, their contents; often tucked away in the seedy, industrialized sections of town; sometimes conveniently cozied up to businesses that rent big yellow moving vans, rental storage facilities cater to our society's lack of space, lack of time, and lack of close, affectionate contacts with our fellow human beings. For some strange reason, people like to pay a small fortune to rent space to store huge amounts of stuff they will eventually throw away. These places are lucrative if you're looking for furnishings, clothing, and housewares.

The dumpsters around these rental storage units are a heck of a lot more sanitary than those in apartment complexes. You might find the discarded remnants of a fast-food meal, wolfed down in the front seat of a Ryder truck. You might even come across a bag of used paper towels from the men's room of the truck rental office. But you won't encounter the kind of rotten refuse common to homes and apartment complexes, where people eat, shit, menstruate, copulate, barf, spit, blow their snot, and change dirty diapers all the livelong day.

Whether these dumpsters are usually secured or unsecured is a matter open to your experience. One of my fans says he's dived these places all over Texas and *never* found a locked dumpster nor a dumpster designated off limits by a fence or sign. He adds that, at some places, the owners were *happy* he was decreasing the amount of trash in the dumpster.

Of course, this particular fan is the size of a small redwood tree. When he gets arrested by the police—not for dumpster diving, but for various bullshit auto infractions—the cops have to "double cuff" him because his chest is so big. I imagine that most places he goes and with most things he does, people are HAPPY that he's doing whatever he is doing and sitting where he WANTS to sit. Different divers have different advantages, not just in terms of location but in terms of their physique and personality. I might add that my experience in Texas was also that folks in that state, on average, are more accepting of dumpster diving than other places.

The feeling about Texas, at least in my case, is mutual. If Texas declares independence from the United States, I will go there and fight for Texas. But my point is that this fan's experience differs from mine. In my experience, rental storage units *frequently* have secured dumpsters, even compactors. But the fact they are places where people throw away a lot of good stuff is not in dispute.

• • • • •

Chad and Robin are a dumpster-diving "poster family." Last time I checked they had four beautiful children, and they are only in their early twenties. They want a huge family and are apparently planning on having as many kids as possible. It's so nice to see young people who have their priorities straight. Robin is a beautiful young woman (with flawless penwomanship, I might add), but Chad is the most striking member of the family. He looks like a movie star.

Robin, the spokesperson for the family, wrote me one of the best letters I've ever received. By a weird coincidence, she wrote it on June 18, my mother's birthday.

Dear John Hoffman—

My husband and I owe a debt of gratitude to you for your outstanding book . . . It is our Bible, our most treasured book, the most effective "self-help" book ever.

I'm not making up this letter. Really. Would the guy who wrote the Bible tell a lie?!

I first heard of you a few years ago when I was flipping through the channels and came across a man happily chowing down on popcorn right out of a dumpster. The next time I came across the name John Hoffman was in the TV TIMES last year, for a 30 minute program titled THE ULTIMATE DIVE. I wish I had recorded it! I watched it and loved it, but had no clue what a force this dumpster diving activity would become in our future lives!

Sometime last summer the Pepsi Corporation had an incentive program called "Pepsi Points." By gathering enough points from Pepsi packages, one could attain any of various Pepsi brand merchandise, including a much coveted (by my husband) Santa Cruz Mountain Dew skateboard. Well, there was no way we were ever going to buy that much Mountain Dew, so Chad and I came up with the novel idea of looking into the recycling bins of all the apartment and condominium complexes for discarded bottles and packages. Not only did we come up with an astounding number of Pepsi Points—enough for the skateboard and hundreds more—we came upon a startling discovery! All kinds of perfectly good household items chucked right in the bin! My scavenger husband would happily gather up all he found that summer and we stored them away for a possible future garage sale. Then we forgot all about it!

Last Christmas I was on Barnes and Noble.com looking for gifts for my husband and there it was! THE ART AND SCIENCE OF DUMPSTER DIVING! Perfect, I thought, and ordered it right up. It arrived four days later and I read the entire thing cover to cover before wrapping it up and popping it under the tree.

Chad loved it! He laughed and schemed and read it aloud. He could hardly wait to get started, he said, he would dumpster dive every single day! However, he said he really didn't think he could ever dumpster dive FOOD. I agreed that it did seem rather sordid and let the topic rest.

A couple of evenings later Chad went out to begin his nightly dumpster visits. I patted him on the head and sent him on his way and went back about my business. Around an hour later came a knock on the door. I peered out and there was Chad standing about ten feet back. I opened the door and there in front of a grinning Chad was a coffee table covered in produce!

Well, the coffee table had to go right back out where it came from, considering its damaged state and unpleasant odor, but I was incredibly interested in looking at this amazing and sudden cornucopia of edibles presenting itself to me. An interesting first experience, none of these perishables were perishing but were merely "visually unpleasing"—oranges with creative overlapping rind, miscolored cucumbers, bags of ready-made spinach salad that weren't even past their expiration.

I immediately took down a platter and began to wash and chop and slice, creating a produce platter to present my children for their breakfast. This was true satisfaction—I knew this was a good thing when I saw it. We were hooked and yet had no idea what a pittance this was compared to the size of what our later finds would be.

To cover only food items—a hundred plus pounds of Blue Bonnet margarine and Land O' Lakes butter, not less than 2,000 strawberry flavor Snackwell's breakfast pastries, Snackwell's everything else in the WORLD in amazing quantities, scads of string cheese (do you know how expensive that stuff is?), hundreds of quarts of Minute Maid orange juice, none of which EVER went bad, no matter how long it took us to drink it, thousands and thousands of cans and single serve bottles of Coca-Cola and all of its offshoots—Sprite, Diet Sprite, Barq's root beer, cherry coke, diet coke, caffeine free diet coke, Surge, et cetera! Cookies, crackers, pasta, gourmet candy, bulk coffee, Coffee-Mate, pretzels, chips, and on and on! Such abundance!

Two of our hugest finds ever were Care-Free sugarless gum—around 30 cases—and 50 cases of imported bottled water in beautiful blue glass bottles. My husband was making good money selling the bottled water to co-workers until one of them discovered that the water was EXPIRED and began to have a complete hysterical fit. I ask you—EXPIRED water? So we save them in anticipation of the next famine and sell them individually to the lovely patrons of our high-class all-dumpster yard sales! We will never rid ourselves of all that cursed gum, no matter how much we sell—any ideas?

(Yes. Little things like that mean a lot to people in jail for, say, protesting corporate evil.)

Now for the fun part—residential dumpsters! It's such an incredibly Zen experience, and a real marriage saver. When we first started, we were very disorganized—now we have it down to a science. My husband goes out and gathers and then brings it all home and dumps it right in the middle of the floor. He proudly demonstrates each item to me just like he's working on the Home Shopping Channel. I either put the items in a keep box or veto them and send them right back out the door, on a case-by-case basis. My job is to clean things up, price them, box them and store them until time for a yard sale.

Other things we keep, such as clothing and toys, shampoo and soap, household cleaners, coloring books—the things we use regularly. We don't keep much—I don't like excess or clutter. I much prefer selling these delightful items to unsuspecting garage salers who certainly have no idea that the merchandise they're tripping over themselves to get is all purely TRASH. It *makes me so nervous and giddy, I nearly get ill from the excitement.*

We only started last January and our first disorganized yard sale in March netted us $600! The last one we had was only a last minute deal at somebody else's sale—and we still made $150 from just a few thrown together boxes! Our goal for our end of August sale is $2,000. This one is going to be a whopper, and we will have no problem making that goal!

Our newest discovery was a used bookstore in our area which regularly gives up on its merchandise and chucks it into a recycling bin. Each morning before work, my husband swings by there and loads up the car. Our next yard sale will have a gargantuan book section—and all really great books!

It's such a pleasure to have so many new magazines to read on a constant basis. It's so gratifying for my husband, when asked where he got his "new" shoes, to say, "I found them!" and then perplex his admirer by admitting that every item on his body was free, and clearly name brand—from Michael Jordan sneakers to his Eddie Bauer shirt, Nike cap and Gap jeans. "What, do you just go through life FREE?" somebody asked him the other day. "Can you think of any better way?" he answered.

I myself am not a clothes horse, but treasure my dumpster Birkenstocks, Helly-Hanson jacket and incredible array of godawful expensive colognes and perfumes. I love my puzzle magazines—usually with only one puzzle scribbled on by the actual subscriber. I adore my little struggling houseplants saved from the jaws of compactor death, cared for tenderly with dumpster "Miracle Gro" and occasionally repotted in dumpster potting soil. I love knowing that I will never again in my life need to purchase shampoo, no matter how many children I have, or how long their hair. I love the "package deals"—the dumpster coffee maker with the dumpster coffee filters, coffee cream and mugs, none of which were even from the same dumpster. I love putting my future garage sale wares into dumpster boxes, marked with dumpster garage sale labels, sealed with dumpster tape and marked "Ready to Sell!" with a dumpster marker.

Even my preschoolers love our lifestyle—a huge box of "new" toys on a regular basis and new books, often "pop-ups" which no mother in her right mind would actually PURCHASE. My children scream out their window, "Hey Papa, are you going dumpster driving?" And their favorite outfits are the ones from the trash. I see people carrying big bags home from Toys R Us and just want to smash them over the head for their stupidity—but it's only a matter of time before those toys are ours.

The other day I had a startling thought—I wonder how many of the things in our home are still being paid for on someone else's VISA bill? Yikes!

I'm sorry for rambling on for so many manic pages, but I had to let you know that you've done a great thing by writing your book and sharing it with others. I've read so many "cut your spending" books and yawned at their futile tips like "use cloth napkins" and "buy an artificial Christmas tree." I don't know about you, but I plan on finding a nice fresh dumpster tree THIS season. An all-dumpster Christmas—I can hardly wait . . . .

I think you're my husband's idol—he'd also love to go dumpster diving with you if you're ever in the area! We hope to hear from you—and thank you again so much! Sincerely . . . .

As it turned out, some time later a reporter in the same region as Chad and Robin wanted to do a story about dumpster diving. Needing a local angle, as reporters usually do, she asked if I knew any divers in that part of the state.

"Hmmmmm," I said. "I seem to recall one. Let me go back and look in my files."

I have the most incredible filing system in half a dozen mismatched dumpster-dived file cabinets, bright and weird with stickers arranged in subversive themes and wandering tribes of loosely organized fridge magnets. (They're not just for fridges anymore!) Files grow and replicate, or merge together, and have odd, idiosyncratic names and characteristics, an evolving "idea and information eco-system" not unlike my own brain. But in an instant I can locate, for example, military discharge papers which might theoretically allow me to apply for a small cash bonus from some state government, or a newspaper column I wrote back in college to prove to some doubter that I coined the term "Americorps."

My orderly and disorderly impulses constantly strive and then cooperate—I am not a piler but a filer—and not a dumper but a diver. Those who visit the abode of the master and expect to find a garbage-filled rats nest are amazed to find . . . chaotic self-organization. So often life is a perfectly balanced, zero-sum game, where the margin between winning and losing life-sustaining energy is tiny. To be able to say, "Yes, you did make such an offer. It was in an e-mail dated such and such, and I've retained it in my files" can be the difference between wealth and poverty, victory and defeat. There is no need to hammer home, too hard, what has been written before, what has been repeated from the luscious lips of Hollywood hostesses.

"As they call it in the book," smiles Leeza Gibbons. *"A Darwinian advantage!"*

Give me leverage and I will move the world. Give me free goodies to sell at rummage sales in addition to my regular job. Give me a window through which I can reach and pull through damaging information about a huge and powerful corporation. But above all, give me the wisdom to organize, organize, *organize* and use well what the universe, in its merciful bounty, throws my way.

Like such a wonderful letter from such a beautiful family as Chad and Robin's . . . .

"How about you look for the letter and call me back tomorrow?" asked the reporter, making an assumption. "Because, to do this story, I really need a local—"

"It won't take me until tomorrow," I answered. "I have all my fan mail in one place, and I can pull the more recent file. Give me just a moment and I'll have it."

The story worked out wonderfully. It was on the front page of the paper—twice! The corporation in question published two different newspapers at that time, which later combined. But what captured and held that reporter's strong interest was not me but the words of Robin, a young woman who, as it turned out, lived only a few miles from the reporter:

"It is our Bible, our most treasured book . . . ."

"Wow!" said the reporter.

"Indeed," I agreed.

Some time later, Robin sent me a delightful card and commented upon our visit slash mutual media interview.

Dear John—

Thank you very much for coming over yesterday and for the new book. (THE SIMPLE LIFE—Berkeley Press, 1998). *I read your section right after you left; it's always nice to have something new to read by someone you admire. You've going to have to clear up this turkey baster thing—I'm afraid it's gone right over my head!*

(Dear sweet, innocent girl . . . .)

We're looking forward to this article coming out on Thursday. I wish we could send copies of it to more people we know. I'm sure my grandmother's head would blow off. She's a product of the Great Depression but prefers to express it by stocking up on lentil soup when there's a sale at Bartell's.

There is a great amount of satisfaction in living this dumpster lifestyle—I know I wouldn't have been able to find a card this great at the Hallmark store. And I love these liquid ink pens! People can't seem to rid themselves of them fast enough. It was also noble of them to donate the stamps. I can't wait for Christmas—sending dumpster cards and dumpster packages. We've got the wrapping paper, several pairs of scissors, rolls of tape, spools of ribbon. We've even got some lovely never-opened dumpster gifts socked away! The JOY of giving will be complicated this year as I try not to shoot eggnog out of my nose watching my sterile Aunt Judy's glee at her "new" stationary. We'll owe all of our Christmas "cheer" to dumpster diving this year with any luck . . . .

I am on the phone now with my friend who works for Amazon.com. She is trying to kill me by telling me how incredibly much her stock is worth and having her absolute zero interest in dumpster diving! Worse than that, she goes out and buys MENTHOL KLEENEX when I have scads I could give her. OK, I'm off. Thank you again and we hope to see you again sometime.

• • • • •

Some divers in Australia invented a novel concept . . . a website that coordinates information about dumpster-diving sites. They also have a website that displays buildings "ready to be squatted" in Sidney, Australia. Bravo, Aussie brothers and sisters in anarchy. Every city of any kind of decent size should have websites with information about dumpsters and buildings ready to be squatted. You can find the dumpster site at <http://www.cat.org.au/skippy> and the squatting site at <http://www.squat.net/shac>.

"Predator," a member of the Cat@lyst Collective, writes the following:

Wow! An e-mail from the dumpstermeister himself! Uh . . . does this qualify, therefore, as junk mail? Heh heh!

Well, I'm honored to be hearing from a true visionary. You have no idea how many kilotons of stuff has been rescued from landfills around the world, consequent to your publication of The Art and Science of Dumpster *Diving. I look around my room and see hi-fi speakers, computer componentry, matching backpacks, books, tools, furnishings, clothing, most of which has been snarfed from roadside throw-outs or from handy municipal, corporate, or construction site garbage skips.*

I think the website is the next logical step, and I'm thrilled you like it. Initially, when Skippy was set up (I think Matt Arnison from Cat@lyst threw the HTML code together one day back in 1997 or thereabouts) we intended it to be a local thing. We were surprised when we noticed that people from around the world were dropping in details of hot dumpsters, but this is a good thing, I think.

Feel free to spread the URLs and encourage people to set up their own dumpster listing system, if they like, they could use ours as a guide. I'm sure the Cat@lyst crew would consider it excellent if this were to occur. I'll bounce it off them tonight at the collective's monthly meeting, and I'm not expecting hoots of opposition to the idea. Who needs e-commerce when you can have FREE commerce . . . no, when you can have d-commerce? No need to tell you for what the d stands, right?

Your sequel, if it is even remotely as good as The Art and Science, *deserves a zillion sales and is a force for positive social change around the world. It should be handed out in schools and used to equip people with lifelong immunity to the later enculturation into our insanely wasteful throw-away society which will bombard them throughout life.*

John, the art you espouse is positively subversive! The pile of money I haven't spent, and the pile of taxes for which I have not been gouged, by failing to purchase the aforementioned and many other things, is something I've failed to quantify over the years, but it would be in the multiple tens of thousands of dollars. If the annual tonnage of dumpster-snarfed goodies around the world was determined, it would amount to something truly enormous, a really big dent in the profits of companies which make disposable, designed-to-break items and the companies which profit from those who pay to dispose of them.

Materially convinced of the brilliant wisdom of dumpster scavenging, many of my friends have taken up the practice, too. Some of them are even recycling things they can't carry home: they squat in abandoned buildings, thereby easing the housing problem and also saving them a pile of money in rent, insurance, etc. Apart from the excellent food to be found, some people are recycling used vegetable oil (or even lard) from dumpsters in fast food outlets and powering their diesel vehicles with it.

There are an estimated 600,000 computers thrown out in Australia each year, and I and others rescue parts from those we find and return them to people in the community. The CD-ROM drive in the computer which hosts our "Skippy" page is itself hot dumpster booty, spotted whilst walking with my dog. Once gained, you never lose your "dumpster eye." Right now I am looking at an SVGA monitor I found in the same dumpster. I take them, test them, and if they don't work I write "broken" on them and replace them in the source dumpster so other divers don't have to carry it home with them only to discover it's broken later . . . .

But what if the bad guys start to do this, mate? Maybe we should add things to the word "broken," something the corporate drones would never write.

Rock on! I think, with your permission, I'll append your email to the Skippy pages, and then anyone who logs in can send you dumpster info from around the world. I imagine it'll start slowly but exponentiate over the years.

Agreed. So it is with the butterfly effect of chaos.

• • • • •

Now I have to tell a story that sort of fits here. Well, OK, it doesn't fit at all but I'll MAKE IT FIT.

BIG BUD'S ESCAPE FROM THE SALVATION ARMY DUMPSTER

Big Bud lives in a small North Dakota town and survives on a small pension. He is a brilliant artist working primarily with materials such as antler, bone, fossilized tusks, and pipestone, and often he engraves glass, though his favorite and preferred medium is moose antler. Bud is a dumpster-diving Santa Claus, constantly scrounging materials and redistributing them to a huge network. He tends to concentrate on building materials and is not overly bookish. He drives a big gray Oldsmobile Regency that would have been a luxury vehicle two decades ago. The trunk of this vehicle is huge; it's literally the biggest car trunk I've ever seen. The back seat is roomy, and Bud further supplements his "dumpster diver's Lexus" by utilizing clip-on roof racks suitable for transporting canoes, skis, etc. The front seat is roomy, too, which is a good thing because Bud weighs about 400 pounds and, like a superhero, goes everywhere with his faithful sidekick, a little black third-hand mutt named Scooter.

Bud has other survival tricks, some of which it is better that I not reveal. Suffice to say, Bud is the only person I know who buys *animal antibiotics* and treats himself rather than screw around with a doctor so he can save gas money and buy himself a *delicious* rack of ribs with home fries at his favorite eatery.

These tricks with veterinary supplies might proliferate, given the current cost of medical care in the land of the free and the home of the uninsured. Hey, it's all antibiotics. It all comes from moldy bread and fruit.

But I digress.

Bud discovered that the dumpster behind the Salvation Army was a fantastic dive spot. Those who drop off goodies at the Salvation Army might suffer with the precious delusion that everything given to the Salvation Army is carefully washed, folded, repaired, put back together, then lovingly placed out for sale to earn money to do the Work of Our Lord. Well, the Lord doesn't have *time* to screw around with most of the crap people drop off, and the most dedicated people drop off the WORST crap, while the BEST crap (like campus crap from wasteful little pukes lacking in life experience and insight) doesn't make it to the Salvation Army most of the time, so if it's incomplete, ripped, or has a picture of a busty babe on the cover, out in the dumpster it goes. Indeed, the Lord's Dumpsters are simply overflowing.

(No, that wasn't a corporate D.)

Now Bud, who is artsy and clever and oriented toward *materials* rather than complete objects, finds much of this stuff useful. One fateful day in 1997, he drove up close to the Salvation Army dumpster, close enough to throw the smaller items directly into his trunk. There was so much good stuff that day that Bud, as big as he is, decided to climb right in, closing the trunk of his car and using it as a stepping stool of sorts. He began pulling out all kinds of stuff, digging himself in deeper and deeper until he had almost cleaned out that dumpster, only to find . . . .

He couldn't get back out again.

Well, thought Bud, this is a bit of a predicament.

Too embarrassed to yell for help, Bud tried to grab something from his pile of goodies to pull back into the dumpster and use as a stepping stool, but everything was just out of reach. Bud thought of tipping the dumpster over, but his car was in the way, and he couldn't very well tip it *sideways* from inside. Bud tried sort of dashing sideways, one end to the other, to push the damned thing sideways, but the wheels were sticky so *that* didn't work either. It was a hot day and Bud was almost dying of thirst. He wanted to cry but couldn't spare the bodily fluids.

Bud realized his only chance to get out of the dumpster was to get one of his feet over the edge and climb out. So he tried. And he tried. And he tried. As he grew more tired he grew less successful. Or so he thought. It wasn't like he could *see* his foot and get any feedback. Bud jokes that he hasn't seen his feet for many years. What a shame that thin and foxy people don't have Bud's humor, generosity, and loving soul.

Carefully, Bud gathered up what little was left in that dumpster—magazines and papers, a mothball-smelling sweater or two—and made a little pile on which to stand, gaining a couple inches. Gathering his energy, putting all his concentration into getting one foot over the side, Bud used his dumpster-diving super powers . . . .

. . . . and got that foot over!

Now it was just a matter of getting out. Heaven only knows whether Bud cursed or prayed or did both, but the 400-pound man precariously balanced on one foot inside the

Big Bud with Alexander James Hoffman.

empty dumpster somehow got himself over the side, almost castrating himself in the process. Bud landed in a pile of books and pieces of plastic crap that broke his fall somewhat but also poked him and hurt him. He just laid upon the pile for a while, laughing at himself and crying at his predicament, until he gathered the strength and will to get up.

Years later, Bud is still laughing at himself. He was in the dumpster of his dreams, finding more and more cool materials, positively cleaning out the dumpster . . . and, too late, he realized he had dug a hole for himself.

"Bud," I told him, a few weeks ago, "I never covered this particular phenomenon in my groundbreaking work on the subject, and I thought I'd covered everything. Do you mind if I use this story?"

That was fine, Bud said.

Just don't use his real name.

And would I like to go share an order of some delicious barbequed ribs?

CHAPTER 6

The Really, Really Weird Chapter

If, god forbid, I were to die tomorrow, my book would continue to change the world. Most people read books without initiating contact with the author, so a single letter from a "Chad and Robin" family tells me there must be hundreds *more* families out there who have been, er, *touched by a diver.*

And yet that's not good enough. There is always more to reach for, more angles and opportunities, and more ways in which the world needs changing. Here are more bits of interesting feedback from other fans going in their own unique directions.

My main interest in dumpster diving (and in getting your book) is in getting and recycling electronics. Perhaps you've heard of BEAM robotics. (That stands for Biology, Electronics, Aesthetics and Mechanics.) Enthusiasts like myself recycle VCRs and calculators into solar powered robots . . .

Mike N., who requests that I write but provides no return address

I get some fairly wild stuff in my fan mail. But sometimes the wildest letters contain the weird seed of a visionary idea or even a glimpse into an actual future as tangible, in its own way, as our own present.

The first fan letter I received was wrapped in tinfoil so "they" couldn't read it, and it apologized for not using the "WASTE postal system." It was signed by "I. Pierce" and, frankly, it scared the hell out of me.

Much later, I came to understand that my allusion to "the real meaning of WASTE" and the character of my dumpster pioneering father, whose name was Willard, struck some folks as in-jokey references to the writings of Thomas Pynchon, a highly acclaimed and mysteriously reclusive cult author who writes about a WASTE postal system—and ways that weird, "tuned in" people communicate at a subtle level—and has a character in his book *Gravity's Rainbow* named Dumpster Willard.

For the record, I'd never read *The Crying of Lot 49* by Thomas Pynchon before writing about "the real meaning of WASTE." I just meant, you know, THINK about it, waste is wasted WEALTH. My father's name *really was Willard,* and it happens to be my middle name, and I can whip out my North Dakota driver's license or my army-issue Geneva Convention card and prove it.

But here is where the weirdness begins . . . possible proof that, unknown to me, my father was the very Dumpster Willard upon which Thomas Pynchon based his minor character, if it were ever possible to pick up such fragile threads and "prove" such things.

Being repeatedly confronted with bizarre accusations that my own dear departed father was nothing more than a smarmy literary homage to Thomas Pynchon, having been told one too many times to read the brilliant works of this reclusive Pynchon, I finally did so. When I came upon the character of Dumpster Willard, I was struck dumb by the coincidence. It was so weird I had to tell Bicycle Bill, who is kind of my personal guru of weirdness.

Even gurus need gurus.

"You know," said Bicycle Bill, "Pynchon used to live right around here in the University District. I read it somewhere . . . in an article, I think."

Being not nearly so organized in his paperwork as me (ironic, since Bill is always promoting order to ME and I am always promoting chaos to HIM), Bill couldn't locate the documentation in order to establish where—indeed, *whether*—Pynchon had really lived in the very neighborhood where I now lived, adding to a mystery which grew more eerie and maddening every moment. I was forced to search microfiche at the library and obtain a short biography of Pynchon off a website. In the middle of all this, a reporter friend called me up out of the blue and began yelling at me.

"I DON'T APPRECIATE YOU WAKING ME UP IN THE MIDDLE OF THE NIGHT AND SENDING ME ALL THE WAY THE HELL OUT TO BELLINGHAM, JOHN!!!" he greeted me.

"Wha?" I asked. "Huh? Who is this?"

"WHO THE HELL DO YOU THINK IT IS?!!!" he answered.

Well, somehow we managed to establish that this call was coming from a reporter named John Shamanski, who had once done a segment called "While You Sleep" about the twisted lives of dumpster divers and youth living on the streets of Seattle. Me and Shamanski had hit it off, and we kept in contact now and then, like when he was trolling for tips about police misconduct. Apparently, somebody had called Shamanski in the middle of the night and convinced him there was a major fire in Bellingham, Washington, a few hours to the north of Seattle. Hell, the whole town was practically on fire, and there had been at least one death. He had gone rushing there, sleep deprived, to cover the story, only to find . . .

. . . nothing. Shamanski went all over town looking for the fire, only to figure out the whole fire business was a damned snipe hunt.

Somehow, in trying to figure out who would play this kind of practical joke, Shamanski came up with John Hoffman the Master Dumpster Diver as the most likely suspect, specifically citing sections in my book about my love of pranks, including a particular instance where I got on the phone and convinced a friend that I was not myself at all but a ruthless collection agent. IF I DIDN'T SEND HIM OUT AFTER THE NONEXISTENT GREAT BELLINGHAM FIRE, WHO THE HELL DID?! (Like I would have a clue about THAT!)

I told John he was wrong. I spoke from the heart, telling him I would never risk the wrath of a valuable media contact over a stupid stunt like that. I was very sorry that he was frustrated and sleep deprived and had been used and abused in this way, but honest to God, I swear, it wasn't me. (This was years before the song "It Wasn't Me" so, well, it really WASN'T.)

Now John began to feel like a total lunatic and an idiot and began to be racked with pangs of extreme self-consciousness. I tried to help him, saying that all the clues he put in front of me—the fact I was in his circle of friends, that I was a bit of a trickster, that the call involved impersonation—it was at least enough for an indictment, but my plea was NOT GUILTY and the evidence of my innocence was NO MOTIVE. And furthermore, in the instance of the prank he was citing from my previous book, didn't I reveal it was a prank before any harm was actually done? So, upon closer examination, it doesn't really fit my *modus operandi*, now does it?

John felt like a terrible heel. I told him that if anybody, anybody in the world, was going to do harm or mischief to my good friend and valuable media contact John Shamanski, I would be the first to call him up (in the morning, of course) and warn him.

So you know, John kind of asked what I was doing and I mentioned how I was researching Thomas Pynchon and trying to determine whether he had ever lived in the University District. And John said, well, there was a great website devoted to Pynchon and I should check it out. Turns out John was actually something of a fan of Pynchon and knew, from the website, that he had once worked at Boeing.

Well, I said—feeling like a dork, because everybody else was getting on the web, all the *smart* people were getting on the web, and I was still using a "shell account" e-mail which couldn't even receive attachments and didn't have enough speed to surf the web—I said, golly, I'd sure love to read that information about Pynchon, but I'm not one of those people who is on the web, but give me a few more months and I will be, by god. And Shamanski, still feeling like a heel for his false accusations, said he would print the pages and mail them to me if we could remain friends. And I said, OF COURSE we were still friends.

There is a weird and creepy ending to this tangent, those of you who do not already know just from reading the word Bellingham.

Bear with me. Let's set the Great Bellingham Fire tangent aside for a moment. Mentally imagine that I am telling you this story over dinner at Mezzaluna, and in order to remember to pick the thread of this tangent up later, I have taken my steak knife and set it up against my water glass at an odd angle so I will see it several moments from now and remember to come back to this tangent.

From the website information, which John Shamanski did indeed mail to me, I discovered that Pynchon had worked as an engineering aide at Boeing in the early 1960s. During that period of time he had written his groundbreaking first novel known simply as *V.*

Now, armed with a firm date, I began searching old Seattle city directories (how farsighted and brilliant of the people who thought to save such things), trying trying trying to pick up the trail of the world's most reclusive author. And, incredibly, I discovered him . . . or rather, the place where he had lived in 1959 . . . only a short walk from my own apartment.

4709 1/2 9th Ave. NE.

"Hmmm," I thought, squinting at the address. "Well, it kind of has a '49' in it, kind of sort of."

I visited the address. I called the current occupant, who had indeed been told by the owner of its previous association with Pynchon but never bothered to read any of his works. I even dived the trash cans but found nothing of value. Walking back down into the street through a recessed stairway carved into the hillside, seeing an oversized V in a nearby sign which read "UniVersity Motors," even noting a mysterious "S with a tail like the Subway Sandwich symbol" spray-painted upon the wall in front of Pynchon's old address, I thought I could perceive small signs and wonders, telltale marks of weirdness visible to the initiated.

"The way you have to go down below to get to *street level*," I thought. "When Pynchon walked down these steps carved into the hillside, as though going *below* the street while really going *to* the street, did he begin to obsess about 'the street and what is below the street,' like his character Pig Bodine? Was there an oversized V in the UniVersity Motors sign back in 1959? Is that what started him thinking about the letter V? And what about the Blue Moon tavern a few blocks from here? Did that inspire the tavern called The Moon in his newest novel, *Mason and Dixon*?"

Contemplating these pieces of Pynchon . . . ordinary parts of my neighborhood which, perhaps and possibly, once upon a time, a master writer whipped into brilliant and enduring works of literature . . . I paid a visit to my dentist at the bottom of my brownstone apartment building. This dentist talked incessantly, drew philosophies of life from the art of dentistry, and went on and on about his departed, domineering father, also a dentist of course.

"Did your father practice in *this same office?"* I asked. "In the late 1950s? Early '60s?"

Yes, he certainly did.

He was almost the only dentist in the U District in those days, and he saw nearly everybody.

Oh my word, I thought. *Could this same father be the inspiration for the dentist in* V *named Eigenvalue?* The father of this same fellow who is, even now, scraping away at my teeth and telling, once again, the same stale joke about the army dentist who used to put all the accumulated plaque upon a Ritz cracker and offer it to his patient at the end of the cleaning?

My father, Willard Hoffman, made an attempt to write a biography of sorts. I gathered up the pages after he died, but they had spilled from the top of the fireplace mantle and were all out of order. Lacking page numbers, they were written upon the backs of forms from a Veteran's Administration hospital. In order to put the pages in order, I had to read each one, written in a trembling hand, and try to find the page that completed the sentence, all the while dealing with their revelations and maddening mysteries.

I put the task aside for almost ten years. But then, after the birth of my son, something began to eat away at me. And I took up the pages again and read them one by one. There were certain things I knew about my father. I knew he was once picked up for vagrancy and became the first person to break out of a brand new jail in Los Angeles in the 1950s. I knew there were only a limited number of people upon whom the characters in *From Here to Eternity* might be based, and my father was a brawling sergeant and chronic womanizer at Schofield Barracks, Hawaii, right before the attack on Pearl Harbor, and he even remembered the author as a quiet, smart fellow who didn't particularly stick out.

Can a human being inspire TWO enduring fictional characters? One a macho, womanizing army sergeant and the other a down-and-out dumpster diver in a novel that almost won the Pulitzer Prize for literature? Are there such characters among us? The protoplasm of fame and art? Carriers of the viruses of world-changing ideas which never break out into fevers themselves but spread the germs everywhere?

I knew certain things about my father, but I was unaware his many travels had taken him to Washington state in the late 1950s and early 1960s, the same period of time when Thomas Pynchon was probably hanging around in the Blue Moon tavern. Then I read Dumpster Willard's words in his incomplete, handwritten attempt at a biography of sorts . . . .

Apple picking time—Portland. Buses line up on Skid Row—one bus has free canteen cup of wine before you start picking apples. He don't mention free meals. 17 cents a box. 18 foot ladder, bag around your neck. (Canvas) First my cup of wine. Now up the ladder. First apple I reached for a big hornet must have liked wine, too—he buzzed like a helicopter right in front of my nose. He stung me, bayoneted is a better word, right on the tip of my nose! From the top of the ladder to the bottom of the ladder backwards, and I took the hornet with me! My nose looked like an apple for a week!

I can't prove my father was the inspiration for Thomas Pynchon's "Dumpster Willard." I can prove, however, that "my Willard" got as far as Portland during the same period of time Pynchon was in Washington state, but there is nothing in his brief, tangential writings or in his stories that I recall about Seattle, not even in his extensive arrest records I obtained from FBI files, J. Edgar Hoover signature stamp and all. (I am still demanding *my* files from those bastards and my request has dragged on for, lo, these many years.)

There are certain forms of weirdness you feel as a tickle at the base of your brain. And things you know subjectively but can't prove. What a *character* he was, and how amusing were his stories! One can see the young Pynchon pouring him a drink at the Blue Moon . . . .

"Tell me more about dumpster diving, Willard . . . ."

And how odd that I should find my way to

Washington state, not even AWARE that I was following in his footsteps, yet again.

In the meantime, I should say, that I have a little son who will never meet his Grandpa Willard in this life yet knows him as a series of vivid stories, at this point told to a three-year-old perspective. Alexander James loves his backpack, so I tell him about how Grandpa Willard was in that terrible war. It was SUCH a terrible war that one time *his backpack was on fire!* Alexander James loves monkeys, so I tell him about how Grandpa Willard had a little monkey, but the monkey ran away and Grandpa Willard was sad, but then *happy* because he knew the monkey wanted to live on his wonderful monkey island with his monkey friends. I tell him about how Grandpa Willard was *silly* and drank a bunch of liquor at a house that belonged to a very important man named General MacArthur, but later he was making sure none of the Japanese were able to shoot the general at a placed called *Leyte Gulf* (Can you say that? LEYTE GULF? I knew that you could!) and *that* was when there was a terrible EXPLOSION and his backpack caught fire! Very involved in the story, my little son clutches his precious backpack a little closer. So involved in the story is he that he cries out, "I DON'T WANT GRANDPA WILLARD TO BE UP IN HEAVEN. I WANT HIM TO BE HERE WITH US RIGHT NOW! I WANT TO TOUCH HIS MONKEY!"

So it is my baby boy begins to learn about the most terrible conflict known to mankind, and how Grandpa Willard—who did all those brave things, fought in a war which was supposedly noble and just—later became an old man who mostly drove a taxi cab and dived in dumpsters and said, "If I had just one wish, I would wish there was no more war."

Those precious handwritten pages about getting stung with a hornet while picking apples . . . what a random and, really, *inconsequential* story . . . photographs of my father taken while he was alive . . . they are precious, more precious than wonderful books and images I might find or even . . . *purchase.*

These things, and our family stories, will be all that my son knows of the grandfather he so resembles. And these things will, hopefully, be passed down until the original characters are more legend than substance. Time and the forward march of procreation has a way of making ancestors of us all. Become somebody's ancestor or become an evolutionary dead end, this is the game so deal with it.

Will we be good ancestors worthy of reverence? Do we consider *seven generations* when we discard something? Be it a can of paint or a photo of a relative whose name we can't remember off the top of our head?

It bothers me when I find discarded family photos, particularly old ones. Personal letters filled with important family history and insights. And you just never know what might resonate as a colorful family legend, prove a link to time, place, or event with a broader history. If I were publishing my science fiction stories, I might write one abut a monkish order of dumpster divers who go about lovingly preserving documents they find discarded which they know will be missing pieces of family history.

One day, a young and handsome monk finds a photo of a beautiful young woman who bears a striking resemblance to other family members in older pictures discarded with the same image of the same young woman. He is so troubled by the way the young woman apparently discarded these photos that the monk feels he must seek her out, explain the importance of such images and their role in illuminating who we are . . . .

You can write the rest of this story.

I once found three of the most wonderful antique photos. Siblings, perhaps even twins . . . are they girls or boys with long hair? One is smiling and the other is grim. What a piece of history this would be to somebody. How many hours would be spent gazing upon it, commenting on the contrast and saying, "You could see the difference even then! And all these years later, that side of the family is serious and the other side is happy-go-lucky. Though, well, you have *exceptions.* But I mean, *generally.* My grandmother told me these two brothers were given two horses that were just alike. They were out riding, and one horse stumbled and had to be . . . put down. Well, OK, *the horse had to be shot.* This was when the one brother became all serious and the other remained happy-go-lucky. The traits of the two families diverged in that moment. I ask you: Who decides these things? What hand of Fate or God?"

And the other image . . . .

Men in masculine hats posing with weapons in front of a building adorned with MORE weapons and the dead, staring heads of deer. How stiffly they pose. How casual they are about their slaughter!

And the third image, this bride and groom. Or are they just engaged? Doesn't she seem a little old for him?

There are certain dealers who purchase such old images. I can see these first two images upon the front and back cover of a rock album, as a kind of artistic statement, rescued from a pile of worthless crap and given immortal life in a new visual context.

Slaughter of the Innocents–the hot new CD by Fraternal Twins!

Well. I see my steak knife standing beside my water glass, and it reminds me I left the thread of a story far, far back . . . .

Ah, yes. Bellingham.

(*No, waiter, no dessert. Just a check, please.*)

Well, my lovely dinner companion, consider, if you will, *petroleum.* I know that I think about it a lot. Heck, I served stateside during that period of time we were fighting a war for the stuff, and all these years, oil has been a heavy weight upon my mind.

(Oh, how kind of you to laugh at such a bad joke. No, you *are delightful. Delightful and sexy. Where is that waiter?)*

Well, consider petroleum. Consider what it *feels like* to be petroleum. First you are a gigantic fern, a forest of such ferns, you are tons and tons of vegetable matter basking, indolent, beneath a primordial sun. It is that sun which is the meaning of your existence, and all day long you spread your green arms in worship. Later, you sink beneath the watery bog, drowning in drink, as it were, plunging deeper and deeper into darkness, subjected to extremes of pressure until you become the black, tarry goo at the very base of all living existence, imprisoned in a rocky purgatory for what seems like, what is, in fact, an eternity.

And then, by an act of reason and will and creativity, some angelic creature beyond your comprehension plunges a pipe, a probe, some phallic symbol of exploration into the rocky abyss and frees you. Somewhere in Midland, Texas, a young woman is impressed by the oil-drilling man and she *puts out* and *puts out* and . . . .

No, do not worry about that part of the story. Consider the petroleum and what it feels like, what it feels like to be *processed,* to feel the fiery and alive part of you pulled free from the tarry parts of you, which become beverage containers, garbage bags, the walls of baby incubators in Kuwait City . . . the fiery and alive part of you, like the dross and straw burning off in Purgatory, leaving only the gold, is subjected to a process having elements of both pain and ecstasy, like *orgasm,* like childbirth . . . one step closer to the stars and the angels it is no longer *crude oil* but *lead-free high octane.* The tiniest spark can unleash, virtually all at once, the awesome power of the primordial sun locked in its essence, but all sparks are kept away, much to the frustration of the gasoline, which seeks a spark as ardently as a lover seeks climax, moving through tubes and pipes and fixtures and valves, moving and moving . . . .

(*Are you OK? You look a little flushed. Here, waiter, cash please.*)

In early June of 1999 a portion of the gasoline is moved through a pipeline owned by Olympic, a *crummy* name for a petroleum pipeline that needs to remain "flame free" if I ever heard one, but I digress within successive layers of digression. The gasoline moves within the pipeline, seeking flame as ardently as souls seek their maker, seeking but not yet finding, *seeking!*

There is a flaw, a break, a bad portion of the pipeline and the gas comes pouring out into the open air, into a body of water called Whatcom Creek. This is apparently not the result of the flaw itself but a mistake made in a control room directing the flow of the fuel, which that day consisted of gasoline but often consisted of diesel and sometimes even *rocket fuel.* Astute readers of Pynchon will recall how important to the plot of *Gravity's Rainbow* was the smell of rocket fuel clinging to the clothing of a certain twisted psychologist, performing bizarre sexual experiments on his own son.

Hundreds of thousands of gallons pour quietly into Whatcom Creek. Cattails wither and

Holand
GRAND FORKS,
N.D.

fish roll up their eyes and die. Petroleum rainbows glisten like prayerful offerings to God. The gasoline is wild, uncontrolled. Somewhere a man in a control room far away detects a problem, a little light blazes up, contained and safe inside a little bulb of glass. At what point this man detects the problem, and what he did about it, or *should* have done about it, whether there was a man at all or a number of men, these facts have to be let out in a careful and controlled manner in government investigations and settlement conferences and so forth . . . the point is the juice is loose, and it is looking everywhere for a flame, for a cigarette or the hot tailpipe of a car, for the pilot light on a water heater if it can pour through a basement window, the desire of the gasoline for flame roars like a dinosaur, like the sound of a prehistoric ginkgo or monkey puzzle pine being torn asunder by storms that raged, shrieked, destroyed in a time before there were *words* to give them the names of women, words to name the very thing the gasoline is seeking, flame flame flame give me *flame* . . . .

Two little boys, ten years old, are playing near Whatcom Creek . . . .

No, I am not making up a horrible story to displease you. I don't want to displease you, my lovely dinner date. This actually happened, and there is *nothing* I can do to change it, but I tell you because there is a strange coincidence, upon which I will support a kind of weird thesis.

(*I have changed my mind, waiter. As much as I wanted to rush off before, me and my lovely companion would like to have dessert. Yes, two of your specials for tonight. I love the idea of* specials *and tend to order them. I mean,* think *about it. On any other night the chocolate indulgence pie is $3.99, but for one* glorious *night you can have it for a mere $1.99! GIVE ME THE SPECIAL, I SAY!!! The universe in its random bounty has decided that* tonight *it will be chocolate pie. I may very well remember this wonderful night for the rest of my life, babies might be born from it, oh it's true, and forever* chocolate pie *will be part of it. God, I say, is good. Of course, if I really* didn't *like chocolate pie, if I had an allergy or some kind of bad association with it, I might look at the menu and find something else, but what have I got against chocolate pie, or mussel salad, or butter chicken? All things being equal, and time being precious, GIVE ME YOUR SPECIAL, and bless you,* bless you *waiter . . . .*)

(*You're laughing again. It is good to see you laugh. But brace yourself, because this is a* terrible *story that will not make you laugh.*)

So then, Wade and Stephen are playing near Whatcom Creek. They are playing with *bottle rockets.* Yes, you may recall that the *rocket* was very integral to the book by Thomas Pynchon. There was a character in that book who bedded many women, but the interesting thing was *where* he bedded those women . . . and he kept a map of his exploits, yes he did . . . within a day, a German V-2 rocket would fall from the sky and explode at that very spot. The intelligence boys had the map of sexual conquests, and had made the weird connection, and all across a war-torn Europe they relentlessly pursued him, a man who might have a clue to the very fabric of space-time, a way to predict the future, such a man might have *tremendous military applications.* You might also recall the odd phone call to a reporter named John Shamanski . . . .

Bellingham is burning. Practically the whole town is on fire. There is death!

The phone call happened almost two years, *almost two years* before those little boys were playing with their bottle rockets by Whatcom Creek. How odd that a conversation about this inexplicable phone call should end with a promise to exchange information about Thomas Pynchon, a man whose writings *hint* at a window into the future, a flaw, or, one might say, an *opportunity for weirdness* in the very fabric of space-time. There are other, anecdotal tales about mysterious phone calls, conversations with prescient strangers that foretell future circumstances, even the Brown/Goldman murders. Consider this, and file it where you will if you can find a place in your gray matter where you can wedge these unsettling facts:

At 10:30 PM on 12 June 1994, a woman called the Wilshire division of the Los Angeles police department, falsely identified herself as being with Channel 4 News, and asked if there had been a double murder reported in West Los Angeles. Sergeant Steve Merrin, who handled the call and recorded it in the duty log, remembered the unusual phrasing used by the woman. He recalled her asking about "sitting" bodies.

At 11:55 PM on 12 June 1994, the murdered bodies of Nicole Brown Simpson and Ronald Goldman were discovered by neighbors near the front gate of Ms. Brown's condominium on Bundy Drive in the West Los Angeles suburb of Brentwood. Mr. Goldman's body was found in a "sitting" position.

There were no other double murders in West Los Angeles that day.[6]

Skeptics doubt these things—hopeless tabloid psychics deserve our undiluted scorn—but I had my own intense conversion experience on this issue about two years after the Shamanski call, an experience with a mentally ill psychic that seemed to tear open the very fabric of space time.

All the truly great psychics are languishing on psychiatric wards and group homes for the chronically mentally ill, eating directly out of garbage cans and living under bridges, facing persecution at the hands of Seattle city attorney Mark Sidran.

These places are where you will find the ones who can predict the future, but hell if you will be able to make sense of it without almost going insane yourself.

The gasoline flows with Whatcom Creek. It flows until it gets near the spot where Stephen and Wade are playing. How precious are bottle rockets to 10-year-olds, in the month of June, rare bottle rockets almost a month before the 4th of July! Did they just casually fire them off, one by one? Or did they "count down" spaceship style, German V-2 rocket style?

Three . . . two . . . *one!*

Hell came to a babbling brook. In the state where Hanford Nuclear Reservation helped to build the first atomic bomb and so many weapons of the Cold War, there was seen a sight that military men, statesmen, and the professional, highly paid individuals who work with nuclear materials had struggled their whole lifetimes to prevent—a mushroom cloud rising in the sky over an American city. The power to Bellingham, Washington, blinked twice.

Not knowing his sibling had just been critically injured, Stephen's big brother George debated with his mother whether a volcano was the source of the explosion, or whether the Georgia-Pacific paper mill had somehow exploded. At Bellingham's St. Joseph medical center, Dr. Cary Kaufman was near the end of a gall bladder operation when a nurse told him about a plume of smoke rising from the middle of the city. Decisively, Kaufman suspended nonemergency surgeries and put the hospital's full surgical and trauma resources into standby mode.

The black plume rose six miles above the city. The explosion heated the creek to 81 degrees, killing every fish in the area. Millions of dollars dedicated to environmental and salmon restoration were lost in the blink of an eye. Fish were found with their heads burned off. Trees were reduced to cinders. The very worms in the soil were incinerated for approximately 1 and 1/4 miles down the creek, including Whatcom Falls Park and some of the area's best swimming holes.

Steven and Wade tried to jump into the creek to put out the fire. It was a 16-year-old who first saw the two boys, "ablaze in a field of black smoke near the creek."

"I saw two yellow things moving and I heard cries," he said.

Why the "things" were yellow I've never seen explained. As an old army medic, I suspect their first layer of skin had been burned off and their adipose tissue was showing. It was 16-year-old Tyrome Francisco who first found Wade, carrying the injured child back to the house. (God, I love the brave, selfless kids of today!) Then he returned and found Stephen, who was *still too hot to touch.* Wade had come running out of the woods, "groping around like he couldn't find his way." Wade, who was even then dying, could only think how he didn't want his mother to know what had happened, thinking she would be "sad and mad" at him.

The boys were taken to the hospital. The father of one of the boys figured out his son had been injured when he rushed home and found wet blankets and an incinerated pair of pants on the lawn. He will never forget the sight of those pants. "I pick up these pants and there's no way you could burn a pair of pants so much." The two boys were so badly injured that their rescuers didn't know whether they should be allowed to see *each other.* Their entire bodies were covered by second and third degree burns. They were in terrible pain. The only areas not burned were their feet—which had been protected by shoes—and small areas around their waists, where their shirts had overlapped with the beltline of their pants.

Kaufman and the other doctors at St. Joseph wrapped them in a cellophane-like substance and gave them enough pain medication to get them through the helicopter flight to Seattle. Their pain, mercifully, was controlled from that point until the very end. Peacefully, they slipped out of this world surrounded by loving family, professionals, and clergy, with prayers and words of love and final blessings.[7]

In the wake of the disaster, the Washington State Parks and Recreation Commission denied an application by Olympic to build a pipeline across three state parks, including the Ginko Petrified Forest. According to the *Seattle Times*, the commission said the timing of the denial was "just a coincidence."

Yes, well, you can chalk up only so much to *coincidence.*

ANECDOTAL EVIDENCE OF PSYCHIC ABILITY IN A SMALL NUMBER OF MENTAL PATIENTS AND GROPING, INCOMPLETE THEORY OF "FUTURE MASS CONSCIOUSNESS" CREATING AN ECHO MOVING BACKWARD IN TIME

(*You are lovely with that chocolate pie on your lips. There. Yes, you got it.*)

The two little boys who—so it has been intelligently theorized—set off the inferno with their bottle rocket may have been unwitting heroes. According to Bellingham city deputy administrator Don Keenan (quoted in the *Seattle Times*), if the gas from the leaking pipeline had drifted farther down Whatcom Creek before it ignited, hundreds or thousands more people could have been killed.

So I have given two examples of odd phone calls jarringly out of place in space-time. And I have yet to tell you about my "conversion experience" involving a mentally ill psychic, or to share what I hope is a reasonably coherent revolutionary theory of space/time/mass consciousness. I would contend this theory is not my own but is *hinted at* time and again in the writings of Thomas Pynchon, et al.

So it would not be "Hoffman's Theory" but "Hoffman's Clarification Of (What Hoffman Believes To Be) Pynchon's Hintings About Space/Time/Consciousness" and it's a shaky, groping, uncertain little clarification at that.

But it's not easy to swim backwards against the very flow of time, so give me a break, huh? How many times have I vetted this theory before delightful and sexy dinner companions, so much so that I have to use an imaginary one just to get from verbal speculation to words on paper, and only *yesterday* the ball bearings went out in the water pump of my Ford Thunderbird while I was upon yet another weird little mission, the possibility of the car being incinerated near Fordville, North Dakota, with me and the four-year-old inside was not so very distant, while the words from the previous pages were waiting patiently back home on my computer screen, the curser which was here ➜ a moment ago blinking tens of thousands of times, waiting, while the universe itself seemed to hurl stumbling blocks in my path as though to say SHUT THE FUCK UP, CHAOS BUTTERFLY DUMPSTER BOY! THE UNIVERSE IS NOT YET READY FOR THIS WEIRD SHIT!

Consider if you will mass consciousness. By this I don't mean things we all supposedly have lurking in our "collective" mind, but rather the *power* of tens of thousands of minds all thinking the same thing at the same time. It doesn't matter if they are thinking BLOCK THAT KICK or I'M A PEPPER, YOU'RE A PEPPER, HE'S A PEPPER, SHE'S A PEPPER, WOULDN'T YOU LIKE TO BE A PEPPER, TOO? The point is that everybody is thinking the same thing at the same time, or over and over again over a span of space and time, and somehow the sum of all this consciousness is much greater than its total. The power of mass consciousness is a *force* like gravity is a force

Let us suppose you had an entire stadium full of people. And, during the halftime show . . . because the regularly scheduled act came down with a terrible head cold . . . a sort of *theorist* gets up in front of the whole stadium and says "Let's try a little experiment in *the power of mass consciousness, shall we?* Please take note of the newspaper I am holding in my hand. What I'd like everybody to do—everybody in this whole stadium to do along with me, if you would, please—is to *concentrate* on this newspaper and make it *burst into flame* with nothing but *the power of our collective consciousness!*"

This little experiment is NOT going to work. But that doesn't mean the underlying theory is invalid. Consider those people in the stadium. Some are going to be interested and try hard to set the newspaper aflame, yes. But quite a few are going to be disinterested, disbelieving, actively hostile, distracted by the need to obtain a hot dog, even mistaken about what the theorist wants done. ("Fire? I thought he said make it go *higher!*") The power of the consciousness collected in that stadium might be compared to the sun's rays shining through a Coke bottle. Sure, you can start a fire with a Coke bottle . . . if you know what you are doing and can use "visual feedback" to focus the sun's energy. Far better to use a magnifying glass.

But suppose you are trying to start a fire from the sun's rays . . . .

Blindfolded.

With wet tinder.

And a Coke bottle.

Even the most skilled Eagle Scout will not be able to start the fire. This doesn't mean the *theory* of focusing the sun's concentrated rays is invalid.

Since I am an eternal optimist, I believe *something* will happen inside that stadium or near that stadium. The experiment will fail, the theorist will leave the stage amid a shower of beer, dropping the newspaper upon the ground. Later, when the stadium is cleaned up, all the beer cans, popcorn boxes, and the very special newspaper upon which *some* members of the crowd were focusing their attention will get thrown in trash bags, which are unceremoniously dumped in the trash. The *famous* rolled-up newspaper which was seen, coast to coast and around the world, and which is even at *that* moment being joked about on David Letterman . . . .

. . . . will unexpectedly burst into flame. Somebody will have to come and put out the dumpster fire. The fire will be chalked up to pranksters and become the butt of yet MORE jokes. In the stadium parking lot, a man argues with his wife . . . she believed in the experiment and *tried* until tears squeezed from her eyes, while he sat there and mocked. In the car, they still argue. For no reason at all, while he is lighting his ten-thousandth cigarette off his vehicle's cigarette lighter, still arguing with his wife . . . he drops the lighter in his lap and burns his leg.

"GOD FUCKING DAMN IT!" he shrieks, and blames his wife for being *incredibly stupid* and causing his accident. And so forth and so forth, tiny little refracted pieces of *mass consciousness,* sparking and smoldering.

I am optimistic and so I believe *something* like that will happen. But you don't need to believe or disbelieve. Just consider the theory for a moment: *mass consciousness has power, but we're not quite sure how to use it or control it.*

Once, I missed my bus and I stood there waiting for the next bus, which was coming in about 20 minutes. I didn't have a book with me to read, nor was there a random stranger nearby with whom I might converse. I was only blocks from Pynchon's old place. I could have walked there and back in time to catch my bus.

I noticed a pink helium balloon tangled up in a tall tree near a gas station. Whimsically, I wished that I might somehow free the balloon. For no particular reason, I decided to try an experiment and willed that I would free the balloon *with nothing but the power of my mind.* I wished and willed and concentrated and began to focus very hard upon the branch in which the balloon was tangled. Finding the place in myself that was trained by drill sergeants in the hot South Carolina sun, I summoned up a near *homicidal rage* against that branch, willing it to break. This I did for almost fifteen minutes. Nothing happened except, a few times, a good stiff breeze whipped up and the balloon strained to be free while the branch swayed.

I felt quite silly, but there was nobody around to witness it.

There was a thunderstorm that night. The next day, at almost exactly the same time (but not too late to catch my bus!), I was in exactly the same spot, and I looked up to see if the balloon was still there.

It was gone. The storm had ripped the branch asunder and freed the balloon.

"Holy shit!" I shouted. *"Holy shit!"*

I have tried, many times, to replicate the results of this experiment. Well, actually, not exactly. I've said to myself, "I broke a tree branch once. Can I break one again?" And I find that I can't, no matter how hard I try. I don't bother to tangle up helium balloons in branches and then try to free them . . . what would be the point of *that?* I "sympathized" with the original pink balloon because it got there by accident. There would be no point in placing a balloon in a tree myself or having somebody else place the balloon there, or to have some misguided person *pretend* as though we just happened upon another trapped balloon, the whole time standing there, broadcasting their *messy* thoughts, thinking, "Let's see if John can get the balloon out of the tree!"

Does mere *consciousness* have power?

In my experience, yes.

So would a *large* and *concentrated* mass of consciousness—hundreds or thousands of minds, even millions or *billions* of minds all thinking along the same lines—would *that* have power?

Error of causality?

(*Who taught you to use such language, my dear? By the way, how is the crust of that pie? You are eating faster than me because I am talking. It looks quite flaky and tasty!*)

Yes, of course I understand errors of causality. That is when the human mind—which is obsessed with *cause and effect,* to the point that experimental subjects who are shown pictures of a man shopping for oranges, followed by oranges spilling to the ground, will "remember" an image of the man stupidly pulling an orange from the bottom of the pile, even though they actually saw no such image—the human mind makes an error of *causality* when it mistakenly connects unconnected things. Dumpster divers are either more prone to this mistake or have greater immunity than the rest of the human race, because everything we do when we dive, all the divelong day, is to deal with unconnected things, struggling to make the connections, with almost zero feedback to tell us when we are right. As Bicycle Bill says, holding up a letterman jacket he found up in Bundy Country: "I figure some guy gave it to his girl, and then when she got tired of him, boom, it went into the dumpster." A likely explanation, but there is no way to prove it (without getting arrested as a door-knocking, weird-telephone-calling, dumpster-diving weirdo in hot pursuit of sorority girls only yards from where Bundy grabbed that poor dear Hawkins girl), and there are a million *other* explanations, including little pumpkin-shaped spirits who place things in dumpsters for the sole purpose of creating confusion in the hearts and minds of dumpster divers.

Here's an error of causality: A week after moving into a new apartment, you go to a new doctor because you have arthritis, aggravated by moving all your worldly possessions. The doctor—who you find very agreeable and professional—writes you a prescription, which works like a charm. But about a week and a half later, you are collecting your mail and an advertisement comes—addressed to you, specifically, at your new address—and the advertisement is for a *copper bracelet* which, the seller claims, works like a charm to fend off arthritis and can be *personally* and *attractively* engraved with your name, allergies . . . why, any pertinent information you would like!

"DAMN THAT DOCTOR!" you think. "He turned right around and sold my name, my private medical information, to a bunch of quacks! I wonder how many pieces of silver he got, the fucking JUDAS!"

Or words to that effect.

But you are, in fact, mistaken. You are making an error of causality. Because if you had waited around for just a moment, maybe peeked in the trash can near the attractive, antique brass mailboxes for the apartment complex, you would have seen that all your neighbors received a similar letter. It is, in fact, your apartment manager who did not sell but simply *gave* the names and addresses of all her tenants to her brother, who is a very nice albeit somewhat weird and visionary man living on a disability, wracked with arthritis, who *swears* by the copper bracelet, and furthermore, the apartment manager had no idea you had arthritis . . . why, she simply knows the apartment complex is full of old people with aches and pains, and her friend the mailman will look the other way while she shoves the letters in the box to save money on postage. (Didn't you look closely enough to see the advertisement *doesn't have a stamp?* From what kind of weird postal system did *this* epistle originate?)

No, dear, I am certainly NOT saying you are old and wracked with aches and pains. You are *erroneously connecting* my made-up example to yourself when, in fact, I just use the "arthritis cure" example because that's the way errors of causality were first explained to me, though I sort of juiced up the story. And I'm glad you brought up errors of causality. I was going to bring them up myself. The mysterious "double murder" phone call, the "John Shamanski" phone call about a fire in Bellingham, the miraculous freeing of the balloon: all these things could be attributed to *errors of causality*. There are a certain number of weird, inexplicable phone calls in the world that might originate with psychiatric patients, and every now and then, one of these "phone pranks of the mentally ill" is going to eerily match up, by pure coincidence and statistics, with an event in the near future such as the Brown/Goldman murders or the Great Bellingham Fire. Certainly a balloon trapped in a tree on a breezy day stands some statistical chance of being freed, whether a notorious cult author is standing there trying, whimsically, to free it with the power of his mind or not.

Believe me, I am more ruthless and critical of myself and these examples than you could ever be. But then, one infamous night at the group home for the mentally ill where I was working (the very place to which I was trying to ride that bus), I had a "paranormal encounter" that blew away all my doubts about errors of causality and seemed to rip open the space/time/ consciousness doorway.

Ask almost anybody who spends years working with the mentally ill and, except for the most cynical and jaded (which rules out about 25 percent right there), you will find they have at least a couple of stories about some kind of strange encounter which led them to believe the following: "While most of the minds of the mentally ill are just kind of broken or messed up, there are a few *very rare* times when they seem to have some kind of psychic power the rest of us do not possess . . . not that it helps them a damned bit, the poor bastards."

(All names have been changed to protect the profoundly insane.)

"Oh, John, I'm so sorry. I'm so sorry your wife left you and moved to North Dakota! You're a good person. You're always kind to me. You don't deserve that! I mean, North Dakota! WHAT THE HELL WAS SHE THINKING? A person can't even make a living out there. It's like Appalachia out on the prairie or something! I'm not trying to be insulting, because I know you are FROM there, but come on!"

Her voice is a nicotine-stained asthmatic wheeze. At night she is troubled by the thought that my co-worker James, a large African-American man, will tie her to the bed and pee all over her body. I find the mentally ill are more easily understood—or more easily categorized and remembered for professional and clinical purposes, at any rate—if you can find a "transformation" myth to explain them. Consider, if you will, Spiderman? Where did HE come from? Well, he was a scientist and he got bitten by a radioactive spider. And Superman? Well, he's not from here. He was born on the planet Krypton, and even though he would be "normal" on Krypton, here his "normality" translates into super powers like the ability to fly on our low-gravity world. And as for Batman . . . .

Well, you get the idea. The mentally ill have their transformation myths, too.

Take wheezy, constantly smoking Darlene Parcheesy, for example. When she was only about twenty-one, a man lured her into the woods, tied her to a tree, raped her over and over, then pissed on her. Darlene tried to deal with matters by not telling her future husband, who she married a short time later. She failed a "typing test" for an important government job by . . . .

"ONE LETTER!" she shrieks. "Oh God, one letter!"

"Look, Darlene," I said calmly on that oh-so-normal day back in 1997. "My wife hasn't left me. I'm not from North Dakota. My wife loves me and we are expecting our first baby. If she were going to leave me she wouldn't go to North Dakota, since we are both from Minnesota. I don't know where you get these funny ideas, Darlene, but you should just chill out and not be troubled by them."

Two years later, my wife would, in fact, leave me and go to North Dakota for educational purposes. To be part of my child's life, I would be forced to follow. Oddly enough, a *Seattle Weekly* article would mention that I was returning to "the motherland" of North Dakota, as though I were from there.

But this was not the weird encounter that clinched it for me.

The ability to *remember* is a gift. And it is hard to remember *insane, unconnected things*. Who bothers to write down, in the daily or even three times daily chart notes, what the mentally ill are *saying?* Who gives a damn about the crazy, unconnected things they are *saying?* But through a lifetime of disciplined dumpster diving, I've developed a gift for recalling crazy, unconnected things and *making* the connections. I can recall what Darlene was saying amid ten thousand other crazy, babbling, word salad conversations because I *remembered* at that moment . . . there was a nutrition program in Grand Forks, North Dakota, and my wife kept *mentioning* it. How it might be cheaper and more cost effective than the masters degree program at the University of Washington. Darlene had just walked up to the front desk and prophesied the future, and I preferred her to go away and smoke herself a cigarette, which she would usually obtain by begging or scrounging butts.

One night I was walking to work, walking down the dark path I had walked a hundred times before, and right where I turn at the cul-de-sac towards the group home, from some 25 yards away I perceived Dick Sousa standing in the weeds near a chain-link fence, staring intensely at the interstate highway. His gaze was so intense and focused that even in the darkness, as I approached, I thought I could turn toward the left and see the *exact spot* where he was staring. I looked toward the spot as well, thinking to myself, what does Dick see? Is there some odd thing he sees upon the highway? Is he contemplating throwing himself in front of a car? Or perhaps he is hallucinating and I will need to tell the nurse via the proper procedure so they can compare it with observations made by other staff. They can sit down and talk to Dick and see if we need to adjust his medication.

Dick was staring at the highway so intently that he was in a place, mentally, where he couldn't hear me approach. Dick is usually a cooperative mental patient, managed easily enough and, like many on again, off again "street characters" and mentally ill people with bipolar mental illness (diagnosed or undiagnosed), he likes to collect many, many buttons and pins and put them all over his jacket, like some member of a modern hunter-gatherer tribe springing up in the urban jungle. If I obtained a button or pin that I didn't want for my own collection (kept in a box, neatly tucked away), I would bring it for Dick as a little gift to be given at medication time, or I would give it to his counselor so she could periodically reward Dick for doing well.

I was chilled, just a little, as I approached. Dick's staring reminded me, oddly, of my own experiment with the tree branch and the trapped balloon.

"Dick!" I yelled, and he jumped, startled. I adjusted my voice, making it soft and gently concerned. "What are you doing, Dick?"

"Nothing!" he said, quickly. "I wasn't peeing."

"No," I said, approaching him. "I didn't think that you were peeing. You just seem to be watching the highway. What do you see?"

"I thought I saw something," he shrugged.

"On the highway?" I asked.

"Maybe it was a little animal or something," he shrugged. "I looked and looked but I couldn't tell what it was."

I turned toward the highway and swept it with my blue gray eyes, eyes capable of hitting 40 out of 40 targets with the M16A1 rifle. The wildlife that managed to make a home in the very midst of Seattle was fascinating to me. The large tracts of grass between and to either side of the interstate were home to many *delicious* animals, including a large number of raccoon and possum. If Dick had seen some interesting sort of animal, I wanted to see it as well.

"Was it on the highway, in the grass, or near the woods?" I asked.

"I—!" he said. "I couldn't tell. I—I saw something move. Looked like an animal or something."

Dick is lying to me, I thought. *I do not know why, but for some reason he is lying to me. I should write a chart note and document this. He might be hallucinating. He might be contemplating suicide. Then again, it is possible he was simply daydreaming, lost in his own private thoughts, which he did not wish to share. Thoughts of having his own car, and money for gasoline, and being able to drive far away from Caisson Hill.*

"It's a little chilly out here," I said. "It's going to rain. Don't you think it's time to go inside?"

I wouldn't have thought much about this incident. Except that the very next day, *upon the same exact spot of the highway* where Dick had been staring, a van packed with a family and all its worldly possessions burned up from, if I recall, a bad water pump. I wouldn't have known about this except that it was in the newspaper and it was later described to me by patients and staff who witnessed the fire and were able to point out the area of the incident which, indeed, had a black scorch mark.

Swallowing hard, thinking to myself that I will explore this weird tangent, I dove into my own private supplies and came up with some attractive buttons. I waited for Dick to appear, but he did not emerge from his room. I finally knocked on his door, gently, and he said, "Come in!"

He was lying on his bed, staring at the ceiling. His room was a little less luxurious than a Marine barracks. Something had come back to me . . . that I once had a nightmare about Dick Sousa. The nightmare was nothing more than the fact Dick Sousa would not stop *staring* at me.

"I didn't see you tonight and I thought I'd bring you these buttons I have for you before lights out," I explained.

"Oh, thank you!" he said, sitting up and taking the buttons. He was as delighted with the buttons as some girls are with new jewelry. "These are nice!"

Perhaps I took too direct an approach. I was pressed for time and eager to inquire.

"Dick," I asked. "What did you think of that fire the other day?"

"I don't know!" he snapped.

"Did you see it?" I asked.

No reply.

"It was . . ." I said. "It was right near that spot where . . ."

"I DON'T KNOW!" he roared, and I jumped.

"I'm sorry," I said. "I didn't mean to upset you."

Dick composed himself.

He put the buttons down in front of me, as though giving them back.

"I don't feel like talking much right now," he explained.

"You don't have to," I said. "It's late. You just keep those buttons, Dick. It makes me happy to give you buttons you like."

"I'm sorry I yelled," he said, looking up from the bed with a "bad dog" expression, watching me with *those eyes.*

"Hey," I said. "I invited myself here, and you are tired. Don't give it another thought, OK? I'll see you in the morning."

"OK!" he said, brightly.

In the doorway, I paused.

"We're friends, right Dick? We get along."

He agreed we were friends. I was always decent to him and went out of my way to bring him nice buttons.

(*Would you care for an after-dessert drink? Yes? Delightful.*)

One does encounter these things. I recall well how my very first "primary" patient, a young inhalant sniffer named Eric Van R. who was on his way to being discharged out of the navy, once casually mentioned the plot of a story I was writing. I had never told him, nor any of my co-workers, about this story. That was my first encounter with this kind of weirdness, and perhaps it was fortunate this happened so early in my medical career. I was not jaded or cynical but still willing to believe the mentally ill might be *special* and not merely broken, filthy people in need of care and control.

A co-worker of mine later told the story of how a mentally ill man named Morris Bellow . . . his "transformation" was that he tried to tie a hang glider to a pickup truck back in 1978 and when he finally awoke from the coma induced by his severe head injury, he was, well, different . . . managed to predict the date and hour of her child's birth. Bellow's reputation for psychic ability was legendary among my co-workers. Nearly every one of them had a story. Morris Bellow wore the same blue knit cap every day of his life and, defying the writings of philosopher Ludwig Wittgenstein, spoke in a private language which we were all compelled to learn, though whether we learned it or not Morris didn't care, and most of it was unbreakable code.[8]

The number "one" was "ana," the word "no" was "not nay," and milk was something we should *never* talk about, as Wittgenstein said, "We must pass it over in silence." Morris would present gifts of elaborately folded, mostly triangular papers which appeared not unlike . . . hang gliders. Upon these papers words, letters or numbers would be written in crayon, pen, pencil, magic marker, whatever. Frequently enough, written upon the paper would be a kind of "answer" or "solution" to a problem, the kind of problem one contemplated quietly, and never mentioned to co-workers, perhaps never mentioned to ANYBODY. Everybody knew Morris was psychic, but it was hard to discuss WHY without revealing our private financial problems, our medical problems, our *sexual* problems.

Before discussing one of Morris Bellow's "psychic incidents," a co-worker might find it necessary to make one or two confessions in a whispered voice. He/she might confess that all morning long, while in the kitchen slinging hash for breakfast, this rather difficult and revealing personal problem was heavy on his/her mind. Fill in the blank with all kinds of things: *Could my sperm motility really be that low, or is the problem with my wife? That girl I talked to on the street but then walked away and the police pulled up to ask what I was doing—was she really a hooker, and is our whole flirty conversation on audio tape? How do I tell my parents what I'm screaming inside to tell them?*

So, while mulling over sperm . . . hookers . . . sexual orientation . . . Morris comes up and hands over a piece of paper while Mr. or Miss Co-worker is giving him "ana pancake" and trying to avoid *looking* at the milk or *stepping toward the milk* and THIS is what the paper says:

LIG Ht 4 ???

And, weirdly enough, this would turn out to be the brilliant solution to the personal problem.

Wait a second! thinks Co-worker Number One. Human sperm weren't "designed" to be exposed to LIGHT in the course of their procreative mission. Could some sperm be light sensitive? Are these motility readings accurate? What if, during this, the FOURTH test, I were to take a few steps to shield the sperm from excess light before handing over the plastic cup? And if that doesn't work, well, I'm convinced. It's me and not my wife.

Wait a second! thinks Co-worker Number Two. It was dark. It was FOUR O'CLOCK in the morning. Even if they have my voice, how would they match it with my face? Even if they had night-vision goggles, I was smoking a cigarette so the LIGHT is going to screw up the image of my face. I'm in the clear!

Wait a second! thinks Co-worker Number Three. What if my mother and father sit down with me and my girlfriend, Heidi? All FOUR of us? And I say, well, I know this is crazy, but I can tell you this if we turn out the LIGHTS. I can't tell you this and see your faces but I can tell you if you'll let the room be dark, and if Heidi can be here with us.

So Morris was sometimes part psychic, part personal advice columnist, but so often his advice seemed to relate to *deep, dark secrets being contemplated right then* that people weren't willing to share the evidence, only the conviction. Incidents like those described above can be chalked up to an error of causality, but the problem was . . . it was just too damned *frequent*.

One day I decided that I was going to systematically collect and document his gifts and see if I could prove "anecdotal evidence of psychic ability" in Morris. This, I thought, could be a *major* contribution to pseudoscience! Of course, the minute I decided to do this . . . Morris stopped giving me gifts. I figured it was one of his "dry spells" and soon he would resume the practice. A whole month went by. Not a single gift proffered at breakfast time over the steam table. One morning, after collecting the gifts of OTHER staff for almost a month, making no progress whatsoever, I quietly gave up on the project in my heart. As I opened the door to begin serving breakfast, there stood Morris Bellow first in line.

He handed me a little piece of paper, grinning weirdly.

"Ay yah," he said, his customary greeting.

And I thought to myself, "The chaotic forces that control the universe have a way of covering their tracks. OK. Fine. Better that I not document his inexplicable abilities, which I just at this moment witnessed *again.* I wouldn't want poor Morris Bellow kidnapped and taken to a secret government facility, anyway."

Incredibly, Morris Bellow is not the best example I've ever witnessed of psychic ability in the mentally ill, though I mention him in passing—and there are *dozens* of other anecdotes I could relate, my own and those of co-workers—but I have chosen the "Parcheesy Divorce and Dakota Prediction" and the "Dick Sousa Flaming Van Incident" because they are clear, dramatic, and firsthand accounts. (Just last night, while these words remain on the screen, a friend gave me a bunch of old LP records. At the top of the stack was SOUSA.)

But I want to emphasize that these encounters are very rare. You might work for years in the field and not have such an encounter. The vast majority of a day spent working with the mentally ill is gritty. It's not like you go to work and start having psychic encounters. Rather, you go to work and first thing somebody is throwing their fucking oatmeal and shrieking about missing cigarettes, and *that* is the kind of crap which fills your shift. These psychic encounters are remembered *in retrospect,* and you are lucky if you happened to chart the words or the incident so you have some kind of confirmation, some kind of documentation, even to *yourself,* to prove that you are not, in fact, losing it . . . confabulating conversations that did not, in fact, occur.

There is a famous female novelist who died by sticking her head in a gas oven, and she has a few cousins who share her last name, and one of them happens to be *crazy as a loon* but not blessed with any literary talent to compensate. Let us call this mental patient Sylvia Bell.

One morning Sylvia walked up to get her medications. As per the procedure, I watched while Sylvia took the pills from her red plastic "mediset" box and I prepared to chart that Sylvia had taken her medication. I must have done this a hundred thousand times in my career, maybe more. There were people in line behind Sylvia and there were people who had been ahead of her. Sometimes they have something to say and sometimes they don't. This is the kind of thing you do over and over, and while the mentally ill are sort of *interesting,* especially when you first encounter them, or even when you first encounter a new patient, it becomes tough to maintain that kind and compassionate mind-set. It is so much easier to say, "Go to bed Darlene, and quit talking bad about James. You know he's not going to pee on you. AND NO THERE IS NOT A CAMERA IN YOUR ROOM AND I DON'T HAVE EXTRA CIGARETTES TO GIVE YOU."

Sylvia looked at the pills in her hand, thoughtfully, and said, "John, would you check something for me?"

"What?" I said, becoming attentive, not wanting a medication error to take place. *The right medicine, at the right time, in the right dosage, to the right person, and by the right path. Check, check, double check. The patient is the last line of defense. If they look at the medication oddly or say something is wrong, you gotta check, you gotta double check . . . .*

"Is there an inhaler in my box?" she asked.

I glanced over at the little wooden spaces where the medisets were stored, labeled with the names of patients, locked up when not in use, the *narcotic* drugs controlled two, three, four different ways.

"Do you even have an inhaler?" I asked.

"No," she answered. "I don't use an inhaler. But I thought I heard *somebody* say something about an inhaler in my box." Sylvia seemed somewhat nervous and agitated, as though an inhaler for asthma in her box was a frightening apparition. "Could you please check? Could you check really good?"

I made a show of looking in her box.

"No inhaler," I shrugged.

"Is it on the counter?" she asked. "Is it back by the nursing station? Could it be in somebody else's box, my inhaler?"

Why is she freaking out? I thought. *This first appeared like a routine inquiry about meds, but now it seems more delusional. Don't feed into the delusion. Don't argue with the delusion. Don't talk about the delusion. You will only reinforce it.*

"I promise," I said. "If anything comes up, I will let you know. But there is nothing like that here, and no reason for you to be troubled by it. Take your meds and have a nice breakfast, OK? It's *bacon* today. You like bacon, don't you?"

She looked at the pills in her hand. Her hand was shaking a little.

"I have a question for you," she said. "I know you are smart, John, and this question is really important to me. Do you think the expression *Mexican standoff* is racist?"

I laughed. A question for the English major and young cult author! Normally I would have indulged myself in this conversation with the cousin of the famous novelist who stuck her head in an oven, but people were waiting in line, shifting back and forth upon the balls of their feet, groaning slightly.

"I wouldn't say it is racist," I answered. "It's a short and convenient expression to describe how two people are paralyzed to inaction by each other's explicit threat, like when two cowboys are both holding guns and neither dares to shoot first because the other would likely get off a shot. If somebody could provide me a substitute phrase without the word *Mexican* in it, I would use it, but I guess I can't think of one off the top of my head. What's up with you today, Sylvia?"

"OK," she muttered, looking at her medication, seemingly fighting for composure. "OK, then could you *please* tell me what *another* phrase means? But I don't think you are going to like the phrase."

"Go ahead and shoot," I said, and laughed at my own joke. "I got you."

"WHAT?" she asked, alarmed.

"It's a pun," I explained. "You know, Mexican standoff? *Go ahead and shoot!* You get it? Never mind."

"Oh," she said. "I get it. That was clever."

She was smiling under a bizarre internal strain, as though desperately trying to please somebody who *literally* had a gun.

"One more question and then you have to go," I said. "This is fun, Miss Bell, but people are waiting in line."

"OK," she said. "What does it mean to say *an impaired Diane and a manipulated James?* Please, John, what does that mean?"

I froze for a second. Here was a patient who knew too much about the weaknesses and foibles of the people in charge. Our nurse, Diane, was in fact subject to frequent allegations of impairment on duty, which she always managed to beat or to cover over. My night shift co-worker, James, was easily manipulated by Diane. They were friends and would sometimes run into each other at nightclubs and dance. James didn't want to cause problems at work. If Diane were ever found to be impaired on duty, in charge of the drugs and the lives of vulnerable mental patients, he would probably want to get her a cab home and save her job. This psych patient appeared to be engaged in the practice of *staff splitting,* basically the practice of deliberately or by accident turning members of the clinical staff against each other.

"OK, Miss Bell," I said, coldly. "I am not going to get into any staff splitting. Take your medication and go have breakfast."

I thought nothing more of the conversation.

A few months later I came to work and found Diane in charge of the drugs, impaired on duty. As I mentioned in Chapter 2, I made a grab for her vehicle keys but got her narcotics keys by mistake. What happened was I had tried to convince her to take a cab home, to get some help. I had tried my entire bag of verbal tricks and yet nothing seemed to work. Give me an insane schizophrenic or a bipolar over a user any day of the week. All the mental energy of a user is concentrated into maintaining and covering for their substance abuse. It is as though they have a little Satanic narcotic demon helping them plot and plan.

Finally, I pretended as though I had given up. It was on her, and if she made mistakes with the medication, she would have to answer for them, not me. *I didn't want to get involved.* And Diane sat there in a pool of harsh yellow lamplight, putting lovely liquid-filled lavender capsules in a red mediset box, saying, "I've ne*fff*er made a mis*hhhh*take."

The keys were on the desk right in front of her, sitting on the desk calendar, hooked to a little leather circle with the letter "D" for Diane. All I had to do was grab those keys. But if I made my play openly, she might very well *fight me* for (what I took to be) the keys to her vehicle as well as the keys that opened the mediset cabinet.

I glanced over toward the sink in the

medication room, where a single clear plastic baggie sat containing a mediset, labeled with the name "SYLVIA BELL" and the next day's date.

"Bell's going home for the weekend?" I inquired, casually.

"Yes," Diane said, defensively. "I took care of it. It's right there." She pointed toward the plastic baggie.

"Cool," I shrugged.

And then a thought came to me. I'm not sure where it came from exactly, and it could very well have been inspired by the conversation with Bell. But it was a thought which came from *complete mastery of my environment.* I knew Caisson Hill like the fascinating lines in my hand, its procedures and subprocedures and exceptions to procedures.

"Doesn't Sylvia have an inhaler that needs to go along with, too?" I asked.

Diane was confused for a second. Inhaler? She didn't *think* Sylvia Bell had an inhaler, unless it was a brand new medical order.

Kris the nursing supervisor had mentioned this issue during Morning Report some weeks back, perhaps even written a note in the shift log. *When clients go home for the weekend, make sure they have everything. Last week somebody forgot so-and-so's inhaler. Make sure they have all the medications which AREN'T in the mediset as well; inhalers, nose drops, Metamucil, etc."*

"I better check," I said. "You know how Kris gets."

I walked up to the cabinet. I made a show of checking for my keys, which I did, in fact, actually have hanging off the loop of my belt.

"Oh, damn it, where did I put my keys?" I muttered.

I turned around quickly and picked up Diane's keys and walked up to the cabinet. I recall the look of surprise on her face, and I could practically read her thoughts: *My hands were full. Oh my God, John just got control of my keys. He has them in his hand. Now I can see his own keys hanging off the loop of his belt! John might be after my keys but I can't SAY anything. I have to play this off.*

I unlocked the cabinet. I made a show of looking inside Sylvia Bell's box.

"Nope," I said. "You're right. No inhaler."

I snapped the lock shut. I pocketed the keys as one might casually and accidentally pocket something, and walked right out the door of the medication room, RUNNING once I was out of her line of sight.

"Jooooohhhhhhhnnnnnn!" Diane wailed behind me, and it reminded me of Lucille Ball from *I Love Lucy.*

I walked at a brisk pace to a back room, where I found James on the computer. I hurled the keys into a drawer and threw some books on top of them, right in front of James.

He wanted to know what the hell I was doing. I produced the heavy ring of my own "shift lead" keys and held them in front of James. We had agreed long ago that whoever was the shift leader for the night made the decisions and assumed the responsibility, taking both the praise and the heat which resulted. We alternated leadership weekly. One week we might do something my way and the next week his way. We generally talked these things out. But when the chips were down and it was something important that could have serious repercussions—whether to call the police with their tendency to use deadly force or to hope the patient would just chill out, whether to authorize the spending of scarce funds for a taxi or to tell the patient, "Hell if you *don't* understand the bus system, you practically live on it. Now you get on the right bus and you get yourself home. Nobody here is sending you a limo"—ultimately the shift leader made the decision.

"I am the shift leader," I said. "I am taking her keys. Do not give her the keys; do not reveal where the keys are. Stay out of it, if you prefer. I am the shift leader. Diane is impaired on duty, *and I am fucking going to handle it."*

"You're going to call her a cab?" James asked quickly.

"Before she has *another* accident and kills somebody this time, damn straight. She is NOT driving home," I answered.

"Cool," James said. "You handle it."

I returned to the medication room, where the confrontation went out of control due to the fact I had inadvertently taken control of her narcotics keys. As I previously mentioned, Diane threatened to frame me for the theft of Ativan. She threatened to tell my wife that I'd had extramarital affairs with various co-workers, including one particularly cute and tasty little muffin who I had (allegedly) converted from lesbianism. While pretending to call a friend for a ride, she called her *lawyer* to the scene, but as it turned out her lawyer was nothing more than a third-year law school student, in way over her head. The fact I managed to write down the lawyer's license plate became "smoking gun" proof of what had occurred. But these facts were not revealed immediately. Diane thought she had slipped out of things, fumbling with what appeared to be DOUBLE MEDISETS CREATED FOR THE PURPOSE OF STEALING THE NARCOTIC JUICE RIGHT OUT OF THE REALLY FUN DRUGS BY ALTERING THEM WITH SALINE SOLUTION AND THEN PULLING SOME KIND OF SWITCHEROO.

Diane was going to be gone for a couple of days. I had about 48 hours to write out a complete and revealing statement about what took place that night, including my reasons for not going immediately to management but rather straight to the office of my lawyer and my labor union. I wrote like a mad man, in the place where one cannot hear the words of a wife, nor even a crying baby. I wrote to save my job and save approximately 60 mental patients from being at the mercy of a nurse who was chronically impaired on duty. I knew that if I didn't move fast against Diane, she was going to move against me. My statement was not short. It looked like a damned novel. And the minute it was finished, I picked up the phone and called work, with Mike Who Reads Only 19th Century Novels on duty.

"The first thing in the morning," I said, "when any manager comes on duty, you need to take them aside and tell them something. This is very important. You need to write all this down and tell them this. *Diane must not return to duty this afternoon."*

When you are in the kind of survival mode where you are fighting for your job, and possibly even the lives of innocent, vulnerable people, it is not customary to sit down and gather your thoughts and compare them with a bizarre, random conversation you had with a profoundly mentally ill psychiatric patient some months ago. It was only after I turned over my written statement and began talking with company officials, my attorney, and my union supervisor that I began to put all the pieces together and I recalled . . . .

Sylvia Bell's strange request that I look for an inhaler in her box.

I thought I heard somebody *say something about an inhaler in my box.*

How did I trick Diane out of the keys? By pretending there might be an inhaler in Sylvia Bell's box. Now my statement about how I gained control of the narcotics keys was being faxed to attorneys and reviewed by the CEO of the multimillion dollar so-called "nonprofit" corporation in charge of the lives of thousands of the mentally ill, a person I'd never even met before who lived at the top of the corporate food chain, whose presence graced Caisson Hill, oh, maybe twice a year.

Do you think the expression "Mexican standoff" is racist?

In my statement, I had to explain why I had taken bold action but then had not come forward to management for two days. It was because Diane and I were having a "Mexican standoff," where my intent to reveal her impairment on duty was effectively countered by her threat to frame me for the theft of drugs and claim that I was involved in extramarital affairs with co-workers. Realizing the people I worked with were "politically correct" and, indeed, not wanting to give anybody offense, I said something in parenthesis about how I was using the expression "Mexican standoff" in the sense of its widely understood meaning, how I was not aware of it being a racist phrase, and how no other phrase really described the situation so accurately.

What does it mean to say "an impaired Diane and a manipulated James?"

I had to explain why I had been alone with Diane, handling this mostly by myself, with James in the back room, uninvolved, and why I had gone out of my way to handle the matter myself and keep James away from the situation. I was forced to explain how I knew James very well, and I knew of his long and friendly relationship with Diane. I was not stating that James would ever deliberately do the wrong thing, but you see, *I know James as well as my own brother.* If I had gone out of my way to involve James in my dealings with Diane, soon I would have been dealing with *an impaired Diane and a manipulated James.*

It was all in front of me in black in white, in my own words, my own unique phrasing. Somehow, psych patient Sylvia Bell had months of foreknowledge of an incident that turned the company upside down. She, the cousin of a novelist, had communicated with me, *another* novelist bringing me my own distinctive phrases from my own written statement *months before that statement was written.*

How did a mysterious caller know about the double murder in Brentwood that night? Where did the call originate?

My guess is that it was placed from a pay phone by somebody with a history of mental illness. The same is probably true of the Shamarski "Bellingham" call. There may very well have been a bored and jaded medical tech only ten feet away while a well-known psych patient got on the pay phone yet again (no, you DON'T stop them from using the pay phone and making crazy calls unless they have lost phone privileges because of calling 911 or unless there is a court order telling them not to have telephonic contact with, say, David Letterman) and, between the smoke breaks around which the mentally ill universe revolves, this mentally ill woman says, "HELLO? Has anybody checked into the double murder? THE ONE IN WEST LOS ANGELES. Who am I? I am a reporter with Channel 4. Yes. What murder? Well, I would think you would know, you are the police. You've got *sitting bodies* where anybody can see and you don't KNOW?"

Over near the front desk, a bored tech shoots a "warning look" at the mental patient. He doesn't know who she is calling, but she is always making crazy calls, and sometimes it gets back to the group home when somebody out there in the real world is feeling harassed. The tech can hear something about "murder." This is a new one: usually she talks about the moon and worries that there are "television and movie powers in my orange juice." The tech makes a whirling motion with his hand, like, "Wrap it the heck up."

AGAIN, I SAY, FUTURE MASS CONSCIOUSNESS CREATING AN ECHO MOVING BACKWARD IN TIME.

When I was a child I had a fascination with abandoned houses. And I thought, well, they are abandoned, right? Can anybody come along and just have them? The building materials, the furniture and books left inside to molder? Why would a kid want to build a tree house? Why not just take my piggy bank and pull out my life savings . . . thirty or so dollars . . . and find out if the man with the necktie from the county government who seems to be concerned about the hazards of old houses with abandoned wells, or perhaps the irresponsible theoretical owners in the Twin Cities, if either one of them will take thirty dollars for the old house? WHICH COULD BE MY CLUBHOUSE AND I COULD KEEP MY TOY GUNS AND COMIC BOOKS AND JUNK THAT I HAVE FOUND THERE! NOT A PRETEND HOUSE BUT A REAL HOUSE AND ALL MINE, MINE, MINE!

As early as the fifth grade I played an imaginary game looking through the car windows. I had to spot abandoned houses. It was my job to document these properties so kids could come along and buy them. When you are a kid you really believe kids should rule the world. That is because cynical magazine marketers write articles in bold, splashy type in kid-oriented magazines with one-word titles, ending in exclamation marks, and foment this immature KIDS SHOULD RUN THE WOOOOORLD revolutionary thinking in order to sell more bubble gum and Backstreet Boys records. But of course it is actually true. Kids should be running the world to a much greater degree than they are right now. But I used to play this game, long ago. I do not know where the game came from, but it was my job to document these abandoned houses.

And now, at the age of 35, I drive obscure country roads of North Dakota on a weird mission with my son in a Thunderbird whose water pump is, as I write these words, being replaced . . . (The shop called a moment ago. They couldn't find the key. Was I sure I had given it to them? I called back. They found it. Sorry about that) . . . and I recall that game I played when I was a kid. WHERE DID THE GAME COME FROM? It is as though a "me" who is 35 "beamed back" the knowledge to a "me" who is only in the fifth grade, the pleasure of the game, of driving around and finding what my child calls "crashers."

"You are going to tell people where the crashers are and they are going to come and fix them up nice?" he inquires, hopefully, dutifully scanning the prairie for clumps of trees that may reveal decrepit farmsteads. There are hundreds, even thousands of them in North Dakota, many forfeited for a pittance in back taxes.

"That is the plan, Alex!" I laugh. "That is why we are driving on these terrible roads with no names, looking for crashers."

"I love crashers!" he laughs. "And I love dragonflies because they eat mosquitoes. Drive around the dragonflies, daddy, don't hit them."

In the future there will be an event that will get a great deal of news coverage, creating "mass consciousness." It hardly matters what this event might be. Yet another government sex scandal. Yet another war that lasts for the space of a sporting season. Yet another global conference met in the streets by brave and beautiful young protesters. The point is that this future event, which has not happened yet, but will . . . is going to create *mass consciousness*. All of those human minds, concentrating together, upon a torn pair of panties. Joyful Koreans, newly united, dancing in the streets.[9] A brave young face standing its ground against a dozen cops who look like menacing Star Wars creatures. The audience for these events will be seeing the same image, hearing the same words, and that will create *mass consciousness*, and this consciousness will be so loud that some people will hear it NOW.

The echo of this consciousness . . . though it is not really an echo at all? An echo comes after. This is the sound which comes BEFORE the sound. It is echo spelled backwards, an *ohce*.

It is pronounced, OH key.

The echo, the ohce of this consciousness, is so loud that it can move *backward* in time, such that the year 2001 (where I am when I write these words) can "hear" a distant rumble from, say, 2032. It works in the other direction, of course, and perhaps more easily. Think of the feeling you get when walking on a battlefield, the feeling of pure creepiness that seems to seep from the ground. But in the case of an echo from the past to the present, feedback is easily available. You can, for example, check to make sure the creepy spot is actually an old battlefield. The past has already happened, so you can't exactly go back and change it.

Or can you?

Some people can hear the *ohce*. They have a moment that feels like a memory of yesterday, or a week ago, or an hour ago, when *somebody* (hard to remember who, exactly) sat down and gossiped about two bodies sitting in plain sight in West Los Angeles and, incredibly, the police had not done anything. They can't even distinguish this moment from an actual conversation with an actual human being, or a "real" memory of something they have read. The knowledge of the *ohce* is simply THERE, and it creates a kind of confusion or longing which compels them to inquire, to comment, to double check. *Is there an inhaler in my box? What about the double murder in West Los Angeles? Good god, Bellingham is burning down, why aren't you guys showing anything on your station?*

Think about certain great moments of mass consciousness. Over the centuries, how many human minds have concentrated upon the image of a stone in front of a tomb, behind which Jesus of Nazareth lies dead? Is the power of this consciousness so great that you can *walk* upon it, like a clear glass floor over the ocean? Is this the faith as small as a grain of mustard which can move mountains?

How do *these words* become part of the equation? By reading about the past incident with the balloon, picturing it, are you in fact "sending mental energy back in time" to accomplish the task? Did the fact that much of the city of Seattle was, in a coordinated fashion, reading a book about a terrible bus crash (wearing buttons to announce they, too, were reading the book so they could converse with random bookish strangers) influence a mental patient to get on Bus 359 and shoot the driver, forcing the bus to plunge off a bridge? Or is it just an "eerie coincidence" that, in a coordinated fashion, thousands of Seattleites were reading that book?

Some people, some of the time, can hear the *ohce*.

That does not mean the future is predestined. I

imagine there may be many possible futures firming up like zygotes quicken up into embryos which quicken up into fetuses and then into crying, red-faced babies. The shriek of the born baby is not merely heard at the moment of its cry, does not merely echo into the future in memory or via a sound recording, but a faint *ohce* can be heard in the past, or better to think of it as the RIGHT NOW and the baby we are talking about will be born in the future.

Or perhaps it won't be. This is how I imagine it might work.

There is a house in a Midwestern city of moderate size, a pleasant enough house with a white picket fence. There is a FOR SALE sign in the yard. Living in the house we find a pleasant enough family, gainfully employed in productive labor, a man and his wife with no children, Jim and Judy. They are both white. The man has a sister named Bernice, who has been mentally ill most of her life. Every month or so Bernice comes for a weekend visit. They play cards, go to a movie, eat at a moderately priced restaurant that is the sister's favorite. She always enjoys these visits and, with medication carefully packed by her group home, they go well enough. Bernice arrives and leaves by taxi.

One month she arrives not by taxi but by van. There is a taxi strike and the group home had to shuttle Bernice over for the weekend visit. Her routine disrupted, she is a little out of sorts. To make matters worse, her favorite restaurant has gone out of business and the only movies being shown at her favorite theater are cartoons and war movies, both of which she can't stand. Jim suggests going to a different theater but Bernice refuses. She is growing more irate and even tearful by the minute, crying out that her taxi driver friend has abandoned her, the Westward Ho has closed its doors forever upon its wonderful smell of porterhouse steak, and to make matters worse, she can't see a movie in the place where she always goes.

"Let's play cards!" Jim suggests. "And we will eat a nice meal right here."

They sit upon the patio. Bernice shuffles the deck over and over. She won't deal, she just keeps shuffling. Her brother sits back and says nothing. Maybe shuffling the deck will make her feel better.

Bernice stares at the wall, growing more and more fixated. She stops shuffling and just stares. The brother is chilled for a moment. A few times in his life he has seen his sister with "demon eyes." One of these times was the incident where he woke up, at the age of 13, and she was cutting herself on a broken window pane "because the voices told me there was an electric bee inside my chest that would kill me if I didn't get him out." That was the incident which caused her to be sent away.

Bernice never came back . . . except to visit.

"What's the matter?" Jim asks, timidly.

"Some people say it was racism, but I think they deserved to die," Bernice answers. "What were they thinking? That they could just move to a little town like this and be white and black and not have people NOTICE? They say you don't have to lock your doors here, but did they think that went for THEM? They should have locked their doors. They were stupid. But how horrible, the way they died, her belly cut wide open, his head at the bottom of the stairs. The bees coming in through the basement window, buzzing in his blood like electricity, bzzzzzzzzzzzzzzzzz!"

"Did you take your medication?" Jim inquires, carefully. "At noon before they gave you a ride here?"

"OF COURSE I TOOK MY MEDICATION!" Bernice shrieks. "I am only talking about the people who lived here! Can't I talk about things in the newspaper? I didn't mean they deserved to die for being white and black, I meant they deserved to die for being *stupid*, but not like that, that was too horrible!"

"I think we need to call her counselor!" Jim's wife Judy says, rushing inside.

"NO!" Bernice shrieks, alarmed. "WHY? THIS IS MY REGULAR VISIT! I HAVEN'T DONE ANYTHING WRONG! DON'T MAKE ME LEAVE!"

She looks around frantically, and her eyes light upon petunias growing by the patio.

"Those are probably the same bees!" she laughs, almost as an aside.

Jim looks at her, horrified, frozen.

"Aren't you going to stop her from calling?" Bernice asks. "I mean, what's wrong with your wife, doesn't she ever read a newspaper?"

"I–!" Jim says. "I think maybe, ah—!"

"THIS IS MY REGULAR VISIT!" she cries. "WHY ARE YOU HAVING HER CALL MY COUNSELOR?! NOOOOOOO!"

Bernice jumps up and tries to get inside. The wife has actually bolted the door. Her husband can handle his sister when she is agitated, but she can't and won't. Unable to gain entry, the agitated mental patient *breaks the window with her bare hands.*

Now one of two things will happen. And these things will have a profound impact upon the future.

Either Officer Burnt will arrive, or Officer Book will. It all depends on who is closer to the address when the call goes out.

Officer Burnt has been on the force for many years. He quickly figures out that this woman is *some kind of fucking mental patient* who should be thrown in the rear of his squad car and driven back to the loony bin with a minimum of paperwork. Yeah, Officer Burnt figures out the story pretty quick. This mental patient, who usually does pretty well on these visits, was agitated because her regular ride wasn't available due to the taxi strike, her favorite restaurant closed, who the hell knows if she got her meds, and she started shrieking some *insane shit,* and who even *cares* what it was about?

It would be kinder and more efficient if people like this were just taken to a quiet meadow and put down like a dog. Burnt would *volunteer* for that duty. He throws her in the rear of his squad car and takes her the hell back to Shady Acres. In response to what she is trying to communicate, he says, "Shut the fuck up, bitch, or I'll pepper spray your crazy ass."

About a month later, Jim and Judy get a bigger home in a different neighborhood. The visits with Bernice resume. She continues to do well, though she never stops missing the porterhouse steak at the Westward Ho. Almost two years later, at the house where the couple used to live, a new family moves in. He is black; she is white. Though this is unusual for the area, it is not unknown, and their lives remain free from harassment. Though they locked the doors at first, after awhile they fall into the habit of not even doing so.

On Adolph Hitler's birthday, a troubled young man walks right into their house and kills the wife in the kitchen, cutting her belly open. She moans and cries out, and her husband comes rushing up from the basement. The young man decapitates him right as he reaches the top of the stairs, then calmly leaves. He is arrested a few weeks later for bragging about what he did, drunk with his friends in a bar.

(BY THESE WORDS I BREAK THE POWER OF MASS CONSCIOUSNESS AND DECLARE THIS WILL NEVER ACTUALLY HAPPEN)

Sorry. We had to add that. Technical matter.

That's what happens if Officer Burnt shows up. Now, the man and his wife will have a story to tell about how their sister *knew,* but it will become nothing more than an obscure footnote. Indeed, they think it wise to keep it quiet. Trite and uncreative people always get around to saying how people who work with the insane, or hang around with them too long, must be insane themselves. If I had a dollar for every time I've heard some variation on that stupid old chestnut! The next time somebody says it I will pull out this book, earmarked to this page, and just point to the appropriate highlighted sentences.

But, in any case, Jim and Judy keep it quiet. In fact, what the brother *remembers* best is his sister's mention of bees, because it reminded him of the terrible incident when the family knew his sister was insane. He can sort of recall something about "belly cut wide open" and "head at the bottom of the stairs," and it troubles him . . . but did he *really* hear that? What if his mind is making it up after the fact? If only he had written it down. He even tries to obtain a copy of the police report for the time Officer Burnt showed up to transport his sister back to the group home. He finds there was no report, just a notation in the radio log.

The chaotic power that runs the universe has a way of covering its tracks.

Let's run the scenario a little differently, shall we? Let's say Officer Book shows up instead of Officer Burnt. This officer is my kind of cop. He is kind, pro-active, a bit of a social worker at times, very bright and intuitive. Somehow he never becomes jaded or cynical; not much, anyway. He enjoys writing long, detailed reports, imagining that one day a cultural anthropologist from the future will find all this useful. His observations of details and nuances are preparation for the day when, he dreams, he will be a detective and solve terrible murders and sex crimes.

When Officer Book shows up at the house, he helps to calm the woman. He talks with her at length, drawing upon classes he took at his own initiative about homelessness and mental illness. She is not cut too badly, and he helps to apply a bandage. There are some cops who stand around and let people bleed to death, waiting for the ambulance crew, but Officer Book knows that members of the public expect a police officer to *do something* and be able to provide basic medical attention. Bernice tells Officer Book all about the murder that happened at this house, and her terrible day which began when her favorite taxi cab driver went on strike, *abandoning* her. Officer Book talks at length about the cab strike, assures Bernice that the taxi driver has not abandoned her, he just needs more money to live like everybody else. The man and his wife are almost *tearful* at the sight of this officer, who is so involved and kind and handles things so well. They resolve to write a letter to his superiors.

Back in his vehicle, the officer writes a relatively long report. It is, at the very least, longer than any other officer would write about such an incident. Officer Book thinks to himself that this report may prove useful to Bernice's counselor back at the group home, with whom he has been in contact on the telephone. Officer Book makes a note to himself to mail the counselor a copy.

Almost two years later, at the house where the couple used to live, a new family moves in. He is black, she is white. Though this is unusual for the area, it is not unknown, and their lives remain free from harassment. Though they locked their doors at first, after awhile they fall into the habit of not even doing so.

Officer Book comes by and knocks. He is helping to organize a neighborhood watch, looking for volunteers. The Smith family invites the officer inside. The wife, Tia, gives the officer coffee, which he drinks with lots of cream. (You'll have to excuse me, dear reader. Just at the moment I was writing this a young man named Craig who lives next door and often mows my lawn entered my house without knocking, scaring the crap out of me, and I'm still not entirely sure about this 13-year-old's true intentions upon entering my house.) Officer Book has a long chat with the family, mentioning by and by that he once came to an emergency call at this house. The family who used to live here had a mentally ill sister, and once she grew agitated and broke that window right THERE, and Officer Book himself helped to bandage her.

The Smiths find this *fascinating!* Why, they saw the old red stain on the patio and *thought* it was blood and weren't sure what to think! How amazing and wonderful to find out! They bought the house from a real estate company and know very little about the family who used to live here. This is where they plan on living for a long time and having children, so such tiny details are interesting. Can the officer remember anything else?

The officer says that he will try, but it was awhile ago. He remembers the woman feeling abandoned because her favorite cab driver was on strike. Remember that? The taxi cab strike? Well, it was before you moved here.

The Smiths want to make sure that being new in the neighborhood won't make it more difficult for them to be part of the neighborhood watch. Officer Book says it will be a chance to meet more neighbors. Tia says they love the neighborhood and are eager to meet more neighbors. Maybe she shouldn't be saying this, but sometimes they don't even lock their doors!

Officer Book takes a long sip of his coffee and then sets it down. He purses his lips as though trying to figure out what to say. And then he speaks from the heart, saying that as an officer he has a different perception of what is *safe*, because he comes to handle incidents and these matters are not always publicized. He agrees this is a nice and safe little town, but locking doors and having secure windows is *always* a good idea. Officer Book isn't quite sure how to say this, but he'll just spit it out: several months ago there was an incident where a Chinese woman married to a white man had eggs thrown at her car and a racist note left beneath the windshield wiper. It was only four blocks from here. Book suggests that we still live in the kind of world where it is best to lock your doors. *But if we get this neighborhood watch thing going, we might be able to prevent incidents like that from happening in the first place.*

Mr. Smith looks at Officer Book and says he appreciates the officer's candor. He is glad Officer Book wasn't afraid to offend, didn't tiptoe around the interracial marriage issue but rather warned them frankly about this incident. Ending racism, Mr. Smith says, isn't just about avoiding offense, it's about being candid and vulnerable with each other. Again, a citizen is almost moved to tears by the open heart of Officer Book. At that point, Tia Smith asks, "Can we go back to that family who used to live here? How dangerous is this mental patient? What if she gets confused and tries to come back to this house?"

Officer Book assures the family the woman is NOT dangerous, but then he thinks for a moment. Something has been coming back, a conversation with Bernice. She was talking about a family, black and white, who died in this very house. Something is odd and tickling at the base of Officer Book's brain.

"I can actually call up the report and check something. Give me a moment, would you?" he says, rising from his chair.

"You can call up the report from your car?" Mr. Smith asks. "That's amazing! I'd like to see that!"

"Well, these are confidential police reports," Officer Book says, with a tight smile. "But let me just go check something, OK?"

Book is able to call up the report and looks at the sentences he wrote almost two years ago, intending to document Bernice's "delusions" so her counselor would have an idea what sparked the incident.

B. seems to sincerely believe that a married couple, "black and white" were murdered at her brother's house, that the wife's belly was cut open and the man's head was found at the bottom of the stairs, with bees buzzing in his blood.

Back at the group home, two years ago, when the police report arrived in the mail the counselor and nurse read it and thought, "Oh, yeah, back to the 'bee buzzing' thing again. She must have cheeked her noon meds that day. We are going to have to put her on the 'cheeking checklist' and really, really watch this."

Officer Book, who is a rational and intellectual man, reads the sentences over and over, always returning to the words "black and white." He is encountering something he can't quite explain, which he wants to attribute to an "error of causality." Yet Officer Book is not without a spiritual and speculative side. Many times, at the end of his shift, he feels that God had a purpose in what happened that day, and God put him in the right spot at the right time.

Officer Book returns inside and takes a huge sip of coffee.

"I reviewed the report," he said. "I wouldn't say this woman is dangerous like a Ted Bundy is dangerous. But if I were you and I lived in this house, I would be very careful about the windows and doors. And if some weird person were trying to gain entry, I would not hesitate to call 911."

The Smith family become the heart and soul of the neighborhood watch and not only do they keep their doors locked, they purchase a home security system. At the appointed day and time, wearing a pair of Desert Storm era pants still bearing traces of egg yolk, a troubled young man marches up to the door of the Smith family holding a Civil War sword behind his back just so. He knocks gently on the door. Mr. Smith takes one look outside and immediately calls 911. The operator urges him to stay on the line, *stay on the line.*

"I've called 911!" Smith yells out the window.

With a shriek of rage, the young man begins hacking at the front door with his sword, splintering the wood, stabbing at the tiny window panes. Within moments, Officer Book careens to a stop in the street. He gets out and draws his service revolver.

"DROP YOUR WEAPON!" he roars, no longer the social worker.

The young man turns toward the officer and begins to march in his direction, sword held high above his head. Contrary to department policy, Officer Book fires a warning shot *into the ground.*

"DROP YOUR WEAPON!" he repeats, but the young man hardly blinks. Contrary to department policy and the kind of wisdom one finds on T-shirts sold at gun shows, Officer

Book shoots the suspect in the leg and he goes down, crying and letting out a shrill, almost female scream.

"Is that her?" Tia Smith asks her husband, looking out the window when things calm down and the ambulance pulls up. "Is that the mental patient lady?"

Later, Book will have to account for his actions before superiors who are fond of him but not as visionary or intellectual as the young officer who dreams of one day making detective. In front of God and everybody, Book states that the first shot was not intended as a warning shot; rather, just as he squeezed the trigger, he realized he had a bad "sight picture" and his bullet was in danger of going into the house, so in that split second he pointed the weapon downward. As he brought the weapon back up, still aware he didn't want a shot to go into the house, he shot the suspect in the leg. At that point he didn't need to keep shooting him.

Book is told "well done" and gets a commendation. Even Officer Burnt concedes the kid didn't do half bad, and they were not in his shoes. Book was at the scene before backup arrived and it was his call to make.

Officer Book will, God willing, one day be in charge of the department. If there are cops like this out on the streets, I hope nobody throws a bottle at them. Officer Book is the kind of policeman who would resign on the spot rather than shoot tear gas at an anti-WTO protester.

Thus the *ohce* which Officer Book detected the day he spoke to Bernice prevented one alternate future from "firming up," and the newspaper headline was rewritten.

The problem with using *ohce* to alter the future is that nobody knows what the future actually is. If you hear the sound of a train, and you look up and see a train barreling down upon you, using a feedback loop between eyes, ears, hands, gas pedal, and so forth you can *get the hell out of the way of the train.* But with the future there is no feedback loop! Are you moving away from the danger or *towards* it? What if putting more secure locks on the doors caused the Smith family to be trapped *inside* with a madman? How can you even recognize *ohce* and distinguish it from pure madness?

And yet I say it is out there, buried in obscure reports, heading on trucks toward the landfill. We can and should harness all our brilliance and will to *literally* change the future for the better. Starting a fire with a Coke bottle, blindfolded, with wet tinder is extraordinarily difficult . . . but not impossible. If the *theory* is right, then a systematic search for *ohce* should turn it up. But we must keep in mind the universe seems to be covering its tracks, or perhaps playing with us. Added to the imperfect Coke bottle, the blindfold, and the wet tinder is the element of an unseen hand which, every so often, seems to *push* our makeshift lens. Is this hand trying to help? Or prevent our progress? Or is it just *playing* with us?

What if so much madness is nothing more than the shifting of potential futures as sensitive dependence on initial conditions causes the shifting of whole *galaxies* as we offer a stranger a cup of coffee? What rainbows are cast by the random refractions of mass consciousness? Is mental illness sometimes a pitfall of individual attempts at enlightenment that harm the physical brain? Or is it a kind of enlightenment achieved? Some of us at the group home came to the conclusion that Morris Bellow was actually an *enlightened being*. If you gave him a gift—for example, a Salvation Army wristwatch at Christmas—he would hand back the gift and, delighted, keep the wrapper, which he would turn into his own "gift" and then rid himself of that, too. His abstention from materialism was profound. He would use the same wax paper cup over and over again . . . *for hot coffee.*

Those who are seeking enlightenment limit their diet, their material possessions, and *meditate* upon something all the livelong day. They may take substances to alter their consciousness. They often chant the same thing over and over. The state of such a seeker of spiritual enlightenment is not unlike that of the mentally ill, who, by accident or volition, go hungry, who usually own very little, and who obsess about the same matters over and over. Their intake of substances such as nicotine, caffeine, and sometimes any other drug they can obtain (depending on whether they are substance abusers as well as mentally ill) is amazing.

If you met the Buddha on the road, would you even *know* Him?

The seeker of enlightenment chants the same thing over and over again. And in our cities, so do the mentally ill.

Change, change, spare some change.

One day it occurred to me . . . those who beg all day are engaged in *worship*. They are worshipping what our *society* worships. If you stand on a street corner for hours and hours and chant about something, is that not *worship?* If prayer and worship is a force, what is the *effect* of all this worship?

My Oriental philosophy teacher, Professor Scott Lowe, who once tried to live in a Buddhist monastery but could not break himself of the habit of ruthlessly killing flying cockroaches that would try to land in his armpits, says the holy men of India are supported by their society because it is believed that by seeking enlightenment "they are engaged in useful work which benefits everybody." For what reasons do people take care of the mentally ill who live on America's streets? For what reasons would we value them at all?

There is no doubt the mentally ill are "messed up." But sometimes it takes something broken to reveal the inner dynamics of what is otherwise inaccessible. And I have come to believe—after a long and paid sojourn into America's "dumpsters of the human mind" which ARE the psychiatric wards, group homes, residential treatment facilities, and unregulated "jungles" occupied by the free-ranging homeless and mentally ill—that there is *something very valuable here hidden beneath the screams, cigarette butts, and spilled oatmeal* which we should, systematically and scientifically, yet also with creative intuition, examine much more closely.

Some of these minds are capable of hearing the *ohce*, and some of them might even be able to break things, set stuff aflame.

In the alternate future where Officer Book saves the day, somebody is reading a headline at the group home where Bernice is living. The address in the newspaper story about the madman with the sword sounds familiar. Isn't that where Bernice used to go for visits with her brother? Wasn't there a police report, ages ago, about the incident?

Even if the counselor digs out the police report from a dusty file in the basement, what does it prove? An odd connection. Bernice said something about "black and white" and the family *almost* murdered was black and white. But they were not *actually* murdered. It's not like Bernice is *psychic* or something. A simple error of causality? One of her many delusions that just *happens* to match up with reality?

Ask Bernice what she thinks of the whole matter and she will say, "Well, they weren't murdered because Officer Book went there to have coffee. Black with lots of cream, tee hee. And he talked about the Chinese woman and the eggs. He didn't want to shoot that boy with the sword, you know. He lied in front of God and EVERYBODY and decided God would forgive him, and later he will tell the truth from his heart when he becomes the chief of police! PLEASE, DO YOU HAVE ANY EXTRA CIGARETTES?"

The chaotic forces which run this universe have a way of covering their tracks. The process of observing the process changes the process being observed. Trying to "prove" or document these things is like trying to catch a puff of smoke in your hands, like finding a needle in a dumpster. Like "proving" that your own father is the inspiration for a fictional character created by the most reclusive author in the world. And yet I believe it is possible. That stadiums full of people willing to cooperate in experiments can be found, and we can discover a way to *focus the mass consciousness.* Otherwise it just seems to happen by accident. But when the crystal-gazing types display a bumper sticker that says, "Visualize World Peace," I think they might just be on to something. In reading this sentence to the end, you will find yourself visualizing John Hoffman with millions of dollars and beautiful women willing to have sex with him, and the power of these words can't be broken, *there—!*

(*Technical matter! Excuse me. Ready to go to my place, dear? No, you are NOT fat from all that food. You look gorgeous. I do believe you would make beautiful babies.*)

For myself, I am utterly convinced in my own heart due to the specificity and detail of the "chaotic footprints" I observed.

Please check. Please. Is there an inhaler in my box?

I thought I heard somebody talking about an inhaler in my box. Do you think "Mexican standoff" is a racist expression? What does it mean to say, "an impaired Diane and a manipulated James?"

It is possible that the power of all the minds reading that paragraph is literally sending it back in time to Caisson Hill, to the mind of a woman whose name is NOT Sylvia Bell?

I think it is quite possible that anybody willing to engage in media manipulation and get access to the *mass consciousness* may not be merely helping their own particular cause but dabbling in the very power which controls the universe. Without even a feedback loop to guide you, I can only suggest . . . be kind. Be good. Be involved and interested in the details of the human race. Be open. And utilize the WASTE postal system in all its many forms.

Well, in answer to a question of long ago.

BEAM robotics? Never heard of it. Sounds interesting. Tell me more, Mike With No Address Who Asks Me To Write Back. Here I am, writing back. Tell me more and perhaps I will hear a story so interesting that the son of my son will still be telling the tale of the solar robots, made from dumpster-dived VCRs.

Let me offer you a drink . . . .

CHAPTER 7

More Mutual Weirdness
"The Fortune Cookie Chapter"

Here's another letter from Mary S., a nice lady whose writing style and focus on certain items reminds me of my dear sweet mother.

Only my deer sweet mother doesn't specialize in mattresses.

I live atop a very rich vein. My 25 story apartment building sees lots of foreign grad students come and go. They leave all their trash in the trash rooms, conveniently served by the freight elevator next to my apartment. I used to just pick small stuff for the church bazaar. But I noticed that the very items offered for sale on the laundry room bulletin board often showed up in the END of the month trash. Priced too high, and the market only exists at the BEGINNING of the month when new people move in.

I more or less specialize in mattresses. I love seeing ads posted on the 20th of the month. "Queen size Simmons, Excellent Condition, $500 new, asking $250." Because I know they're only advertising my next month's stock. In June and September, I can sell a few before my free ad on the bulletin board needs to be replaced. (Every two weeks.)

I've read books on marketing to help with pricing and ambience. My line for "Where did you get this mattress?" (each customer sees only one) is "a friend had to move before she could sell it, so I'm selling it for her." 900 apartments. As many people move in as move out. No sales or income tax. I'd love to see a franchise of my operation in every building in the country, except mine. It's hard to buy any lower than I do. I set my price just low enough that buying isn't a big decision.

Apartment complexes are indeed lucrative when you're inside "intercepting" waste, which is what this woman is doing. After my first book was published, I was the assistant manager for a 124-unit high rise in Seattle, and I made out like a bandit intercepting good stuff bound for the trash and keeping on top of the goodies that actually made their way to the dumpster, where I frequently had business anyway. (It's nice being PAID to dumpster dive as part of your straight job.)

I would tell people to put good stuff in a pile in the corner and I would "give it to charity or use it myself, not that I can think of anything I need," and, of course, I wouldn't charge to haul it away. For people "burdened" with excess furniture, this was a godsend. I would use space in empty apartments or storage areas until I could turn the item around. I knew better than anybody who was moving and who needed what. Sometimes I "owned" an item for only a few minutes. Once, I was helping the new owner of a couch haul it upstairs when we discovered the fabric on the back of the couch was missing. (And it sure looked like a *great* place to hide things!) He looked with dismay at the couch he had purchased, verbally, for $20.

"Damn," I said. "I didn't notice that."

"You *didn't?"* he asked, peeved.

"I just *got* the damned thing," I answered. "Tell you what. How about I knock off five bucks? And look, it's a great place to hide things. Not that you have anything to hide."

And so, having renegotiated our deal, we continued up the stairs. Another thing I did was establish a "free box" in the laundry room. Lots of people benefited from this box, but I think my family benefited the most. We even obtained a miniature television set which my wife took to work . . . but not when The Big Boss was around. Best of all were the items I offered to take from tenant's freezers, as a favor. Frozen steaks, shrimp, Popsicles. It was a lucrative niche, and we ate like kings.

Let's continue with Mary's letter . . . .

My husband claims I think of picking up mattresses the way other people think of picking up a few $20 bills. He's right. He's also the perfect size to be clothed in Chinese students' castoffs. 32-29. He loves the tags that say "Size XL."

Yes, here's a good time to insert something. Students from Asia are wasteful as hell. American college students are shamefully bad, but these visiting students are WORSE. My theory is they don't have the connections to efficiently sell, give away, or barter these items in a strange country, and in their own country they are no more wasteful than American students, relative to their own country. (Though the tales I hear about Japanese "gomi" diving tell me there might be a nation out there even more wasteful than the United States.) Considering all the food I ate in Seattle from fleeing foreign students, well, in the words of a song, "I think I'm turning Japanese, I think I'm turning Japanese, I really think so."

I don't barter on the open market, but do indulge in a fair amount of mutual back scratching. This morning I told a friend I wanted a trapeze coat to wear to Paris. Could she believe how expensive patterns have gotten? She wants to get into stir fry. Would a skillet work? She has several swing coat patterns from Vogue. Since everyone in my building leaves a rice cooker and wok in the trash room with their mattresses, I often sell them as sets as well. But I'm certainly going to offer my friend her choice of circular, stainless, or carbon steel when she brings those patterns on Sunday.

This morning I woke up on my trash-picked Sealy-Posturpedic under my trash-picked down comforter. I got up, used trash-picked soap, shampoo, conditioner, razor blades, shaving cream, towels, and toothpaste. I climbed into my trash-picked underpants (Jockey for Her—Australia) and knit tunic (Britches of Georgetown). Actually, I met my husband at a used bookstore. He claims I trash-picked him as well. I paid seventy-five cents for my tights at a church bazaar—I'm sometimes willing to spend a little for exactly what I want. Bill slips on his trash-picked Armanis or Calvins.

Yeah, I loved your book. My favorite new idea was reading more different magazines. I found the Grad Program at the Agriculture Department has a free foreign film series.

This would be an excellent time to slip in a relatively recent Department of Agriculture statistic: 25 percent of all food in the United States ends up discarded, and that does NOT count "crop waste" in the fields. Responding, in part, to these kinds of facts, President Bill Clinton . . . and I'm not going to crack one "trashy" joke about this man, mostly because I can't spare all the space, but if his *mother* didn't do some dumpster diving in her life I will eat my shoe . . . signed Public Law 104-210, also known as the "Good Samaritan Bill." Unlike another Good Samaritan law, this one has nothing to do with car accidents. What it does, basically, is to protect businesses from liability (lawsuits) when they donate food bound for the dumpster to a soup kitchen or food bank, as long as they practice normal and reasonable care. (For example, don't donate tainted meat when there is ample documentation it is tainted.)

Ever since this law was signed by Clinton on October 10, 1997, it has been publicized around Christmas and Thanksgiving every year, reminding businesses to donate. Despite this bill, "fear of lawsuits" is still cited as a reason for being lazy and slovenly and throwing away perfectly good food instead of keeping our country strong by donating it to a good cause. But pardon me while I take some credit for what little progress these kinds of half-assed reformist efforts do make . . . and half-assed reform is better than nothing at all, in my opinion.

This is the evolutionary result of *dumpster mass consciousness*, an attempt by the system to respond to feedback, a response on a *national* scale to all those articles, radio interviews, videos, and particularly the unsung efforts of heroic divers in every city and town, crying out against such wanton waste and demanding reforms to anybody who will listen.

We'll get to those stars yet, oh, yes we will. We will become like the angels, living on light, and we'll pull our way to the stars and the angels one dumpster at a time. Meanwhile, though I loathe becoming a *political moderate* by suggesting more and better half-assed reforms, I will have to swallow my bile, consider the greater good of the planet, and do just that very thing.

FIRST, 104-210 isn't working very well. This federal law is far removed and not as immediate and worrisome as, say, the various and sundry state health departments that your friendly store or restaurant manager must deal with constantly, and—in my considerable experience and observation—this law is having a good but SEVERELY LIMITED impact as far as keeping usable food from being discarded.

Restaurant and store owners must be sought out, individually, enlightened about 104-210, and issued appropriate and informative documents explaining how it supercedes particular state laws and agencies in their state. This is what needs to be done to stop the food from being wasted, not merely talking at them in a reassuring Clintonian honey voice upon the radio at Christmas and Thanksgiving.

SECOND, there needs to be a "munificent manager" law in addition to the Good Samaritan law. This would be a law saying if a restaurant owner, store manager, etc., allows an *employee or other party* to salvage food "designated to be discarded," he or she is protected from liability just as though the food were donated to a homeless shelter or soup kitchen under 104-210.

This law will NOT deprive soup kitchens or shelters of donations. The employees who are frugal scroungers are likely to share their windfall quite widely, because NOBODY can eat as much waste as we are generating. Imagine if every American had to eat 25 percent more food every day for the rest of his or her life!

No, this reform will be a *good* thing. Business managers are more likely to trust an employee not to sue them over bad chow mien than they are likely to trust, let us say, a person from the food bank asking for donations under Public Law 104-210. The manager is likely to consider the food donation to the employee a "favor," which may be rewarded with hard work and loyalty, rather than "charity" for some unknown person who offends his anal-retentive Protestant work ethic. ("He who will not work, neither shall he be allowed to eat . . . *from the trash!*" No, this would be too tough even for Captain Miles Standish.) The employee has the advantage of being inside the business and right on top of the source of waste and will personally benefit, so he or she will work hard to assure the capture of that waste.

THIRD, we must *legalize and encourage dumpster diving*, everywhere and all the time.

Period.

My son was born 6-11-97. How odd that in 1994 I would see, scrawled upon a dumpster near a Furr's Supermarket in El Paso that I was known to frequent, the words "Planetary Law 6-11-97."

By the authority vested in me through the Dumpster Goddess, I hereby declare Planetary Law 6-11-97 to *legalize all dumpster diving in perpetuity*, and I order all nations and authorities upon the face of the earth to appropriately conform their laws, rules, policies, and procedures to be in compliance. As part and parcel of PL 61197, I declare all trash compactors and locked refuse disposal areas to be lawless, and I order their immediate destruction and/or liberation.

Further, all individuals imprisoned for trespass and/or other crimes solely pertaining to and stemming from the diving of dumpsters are hereby ordered released, their dumpster-dived properties returned and/or appropriately compensated, and reparations paid for their suffering.

So mote it be.

Now back to Mary's nice letter!

A new one for you, John—I haven't bought laundry soap since I moved in. Enough (residents) leave a half cup in those anti-drip plastic containers to meet my needs nicely.

Not a new one for me, Mary, but a good trick worth publicizing. It helps to throw a little water in the plastic bottle and swish it around, too. Also, salvaged fabric softener sheets can be used more than once, or doubled up once they get weak—two used ones instead of one new one, etc. But now I sound like Amy D. of *The Tightwad Gazette* and it disturbs me, so I'll stop. What kind of woman goes around talking about tight *wad, tight* wad, *her tight little wad* all the divelong day?

This is a woman with issues.

Suffice to say I'm always on the lookout for tricks to cheaply bring me more energy, more life force. Always I think of myself as an "evolving super being," and each new element in my life is part of that evolution. So if I manage to scam free cable, for example, I think of that cable as a tendril of my "super organism," which includes not just my body and mind but what I control and influence, always judged by flows of life force across space and time. When I save a dollar, I have saved blood cells and flesh and pain from leaking into the cold, dark, lifeless void and eked my way up the ladder toward godhood. When I establish a system, like free cable, which saves me money EVERY MONTH, it is like I have grown an appendage, a giant sucking tendril plugged into a stream of life energy. And it is this outlook that really gets me DAMNED EXCITED ABOUT A TRICK TO SAVE MONEY ON DETERGENT BY SHAKING IT OUT OF SOMEBODY'S DISCARDED BOTTLE!

In closing, Mary writes, *We love making up stories about people whose trash we find. What life event leads a person to suddenly discard 9 umbrellas, 6 nearly full gallons of gin (different brands), and 200 colored pencils?*

Best guess—she evicted a drunken English artist boyfriend.

That's pretty good, Mary. And the way dumpster divers can *think* like that, and *learn* to think like that, makes them very creative people. It might be complete bullshit, but it's very *creative* bullshit. How well I recall dives with my brother, but especially the post-dive sorting, which turned into zany improvisation and comic speculation. And if you run out of laughs, there is always the ol' turkey baster.

Here's some great stuff from Timothy M. in Michigan.

I have been dumpster diving my whole life and have stores of stories which people just can't believe. My wife and I have college degrees from a small private college in southern Michigan (honors, etc.), and I am finishing a Masters in History. Of course one of my favorite dives is the university library, where those entrusted with the repositories of great and dull data do periodically take it upon themselves to discard reams of materials for absolutely no good reason other than it committed the crime of taking up space. The purpose of this letter is to inform you of one such dive that you may take joy in knowing that your encouraging tome assisted in helping the following drama to unfold.

Our university was getting ready for its annual book sale (you bet I dive that dumpster when it's over baby!!). Whenever some mid-level library bureaucrat decided that an object wouldn't sell, he put it with the stuff to be dumped. Naturally I was in the dumpster when Miss Library Student Worker came to dump another five boxes of goodies.

"Are you looking for boxes?" she politely asked.

"I noticed this map, and then saw some others."

"There were two boxes containing almost every National Geographic map published since 1940. I LOVE MAPS.

"Oh. Well we can't sell all this stuff (her dumb boss told her) so we have to chuck it out (said the sheep). Here's some more."

Timothy, that's what I call "service with a smile."

"O.K." said the wise and experienced dumpster diver, trying to control his enthusiasm. Among the maps and books and old copies of CIVIL WAR TIMES (!) was a mailing tube 20 inches long and 3 inches wide. Peeking in the end I saw there was something in it. As is right and good, I grabbed it to be inspected later. I walked with my loot to my job in the cafeteria (who wouldn't want to make preemptive strikes on cafeteria discards? Hee hee. Two days ago I scored an entire pan of delicious bread pudding because, oh no, someone left it out longer than some dickinasuit in a Washington D.C. OSHA office said was "safe." OCHH NOO GEEERMS BOOGA BOOGA!) Where was I?

The tube, the tube! What was in the tube!?

Ahhhh, the tuuuube. I went to my locker and ditched my cache. But before I put the tube in my locker I grabbed the end and carefully slid out the roll. The blue paper unrolled about a foot when I was able to see a facsimile of some old book's title page. Then I saw the very edge appear of old paper, old paper which was tucked inside the blue paper. Scholars like us recognize that high rag content of really old books. I unrolled some more of it and HOLY MOSES IT WAS A PAGE FROM THE BOOK OF MACCABEES, AND IT LOOKED REALLY OLD!!! I looked back at the blue facsimile page . . . .

"BIBLIE BREVES INEADEM Annotationes, ex docifs inter pretaionibus, & Habraeo rum commen tariis MD . . . ."

"MD?" I thought. "MD? Whoa. That's the year 1500."

It was a genuine page from the Latin Vulgate of 1500.

Stunned silence.

Then I fumed at them in every fiber of my being. I stomped my feet. I gnashed my teeth. I sang. I rejoiced. I praised God in Heaven. When I got home . . . (I sat down with my wife) . . . slowly unrolled it and discussed my find. "What else is in the tube?" she asked. NO! It can't be. There were more. The total haul was the page mentioned, another Vulgate page from 1700, two from 1600, and a page from the Book of Daniel, printed in 1490. That one page is 507 years old. It was being read before Columbus discovered the New World. It survived the Reformation, the Thirty Years War, and scores of (other wars), including at least five major European conflicts in the nineteenth century and two World Wars in the twentieth. Fires, flood, pestilence, stupidity, and countless other factors pitted against it and bet upon its being trashed into the past like the majority of those items like it.

But no.

It survived.

It survived (as part of a book) until some benevolent doofus tore it out and, on December 20, 1970 (the tube is postmarked), stuck it and others like it and mailed it to my soon-to-be alma mater.

"Frame these and enjoy them," wrote the doofus. "They are free for you and other schools! Let me know if you want more, I'll send you the complete set!!" stated the enclosed letter.

But they could not survive thirty years (at the school gifted by the doofus) and THEY WERE IN THE DUMPSTER! THE DUMPSTER, JOHN!! They would be ashes by now, or rat bedding. But THEY ARE NOT! I SAVED THEM FROM THAT TERRIBLE FATE!! TIM THE DUMPSTER DIVING HISTORIAN HAS SAVED THE ANCIENT HOLY MANUSCRIPTS FROM PERDITION!!

All have certificates of authenticity, too.

I was telling the folks how important that is, Tim.

It shouldn't surprise me anymore, but it does. Do I offer them back to the library? Do I ask the archivist if HE knows what his idiot co-workers are doing upstairs? Do I make a big showy donation of them with a banquet in my honor and then tip my hand during my speech?

Golly, Tim, that last one sure appeals to the showbiz side of me! But I'm thinking eBay, eBay, eBay. Enjoy that stuff for a while, show your friends, tell the story, but then SELL THAT STUFF TO SUPPLEMENT YOUR INCOME, which won't be much as a teacher of history.

Do I have the newspapers come and take my picture next to the dumpster with the tube of treasures?

Only when you get tenure, Tim.

Who could argue AGAINST dumpster diving now?

Those who serve the devil, Tim. And there are a lot of them, Satan's little middle management helpers and the Luciferian manufacturers of compactors. Raise the lid, I say, and purify with God's Light!

Or do I go for a combination of options? Do I let the head librarian find out by reading the paper? Do I ask to see all trash on a daily basis before it gets put outside? Do I DEMAND it? What if the Fat Cat Alumni Donors found out that this is how the school handles its resources?

Good heavens, Tim. When I related a similar story in *TAandSoDD*, I thought it was a FLUKE, not a national trend. Now I'm beginning to wonder. The fact you work for the university in question AND go to school there really complicates matters. How shall I advise you? What am I, Dear Abby (Hoffman) of the dumpster?

Plus, you haven't even mentioned the possibility of seeking out the rich doofus gifting the world, unsolicited, with pages torn out of old holy books. Maybe he is loaded and can give you a big grant to do your own private research. (*Wild West Burlesque: What Elements Remain In Its Modern Day Equivalents?*) All I can say is, you did the right thing by telling me so I can share this story. And isn't it lovely to hold those old treasures and contemplate this unique problem which, like most problems, is really an OPPORTUNITY?!

Dumpster divers constantly encounter complicated "resource" problems like this and must seek out the best solutions. There are plenty of folks who would love to have your problems, Tim M. from Michigan, who has now earned the title "Indiana Jones of the Dumpster Divers" from his guru.

I hope my writings about dumpster diving are helping to fill the world with problems like these.

Post Post Script: Remind me to tell you how I used the trash for valuable intel to get some snotty frat boy to crawl into my room and plead for mercy concerning some minor grievance I had with him! It was Psych Ops at its best!

So I wrote to Tim, congratulated him, and let him know that I'd mentioned the recovery of the antique documents during my weekly show in the Republic of South Africa on the SAFM radio network. And he wrote back and told me this story:

O.K. How I used the trash to make the frat boy beg for mercy.

One day I returned to my college dorm room (I was a Resident Advisor, a cool one, mind you, and since it was the early afternoon I climbed into bed for a post-lunch snooze. All of a sudden OUCH!! What the OOOOWWW!! My elbow got SLICED on something. Bleeding all over the place, I noticed several very small glass fragments in my bed. Frenetic, wounded, panicked, I seethed as I bandaged my arm.

Now, there were some rather irritating freshmen in the dorm, and these ex-con wannabes would break stuff, steal, and raise hell at 3 AM, puke on the carpets, etc.

And waste perfectly good stuff! But it's in my other book, Tim.

Upon careful examination and reconstruction of the glass, I concluded that it must be an ampule. Inspection of the wall revealed a tiny pyramid of broken glass powder, a point of impact if I ever saw one. My door lock was secure, leaving the chained bathroom between my room and the next as the only possible source of entry. Someone who had access to that room came as far as the bathroom, opened my door as far as the chain would allow, threw an ampule against the wall and thus littered my bed with broken glass fragments. The ampule, I discovered, wasn't full of morphine but our other good ampule friend, the ammonia sulfate "stinkenbomman" from Germany! My rightful target was the science major, Jim, a normally quiet and reserved fellow who didn't seem capable of such a thing, but all signs pointed to Jim.

Oh God, I'd be so pissed. And I'd be wanting some dirt on Jim, and I'd go straight to his trash. Not that I'm a "vengeance freak," as I wrote in my other book. But the guy made you BLEED, and you apparently didn't do anything to deserve it, and he's a little sneak and a weasel who misapplies science . . . .

Every night for a week I snuck down the hall at 3 AM and nailed the big bag of collected garbage from our floor's garbage repository, placed there by those cleaning personnel who cleaned the rooms. I had hoped to find evidence such as another ampule, or maybe the blue box in which they come.

Nothing. That is, until the day when there was SOMETHING.

It turns out our freshman Jim pined to become a member of "I Eata Thigh" Men's Fraternity, the wealthiest and most prestigious frat on campus. As a fledgling pledgling he was required to memorize a certain secret oath explaining all about the frat's ideals and why one such as he would desire to become a member. Jim found that handwriting this document over and over was a helpful memorization aid. When he was done he discarded the hand-copied secret oath into the trash.

Oh, now you got him. That's enough to fuck up his life. So what did you do? Did you give the secret oath to the frat's arch enemy along with clues that led straight back to Jim? Something like that?

Well, I knew a few of those frat guys. I'd attended some parties, made friends, personally got them out of a couple serious scrapes and jams through my contacts with central administration. So I sent an anonymous letter from "some of the frat guys."

But not REALLY. I get it.

The letter went something like, "Jim, it has come to our attention that you have information about a prank played on our friend Tim, blah blah blah." It went on to state, "Tim has saved our asses a number of times" and, most importantly, LISTED THE SPECIFIC ATTRIBUTES OF THE SECRET OATH, saying how it wasn't very "kind, gracious, loyal" to pull dangerous stunts that make proven friends of the frat bleed.

Oh, I get it. Naturally, Jim is going to believe with all his heart that the letter is *really* from the frat. Who else would know the secret oath? Well, what did you make him do? Humiliate himself in some public way? Your own personal "hell week within hell week?"

The letter told Jim that his good standing was in jeopardy and that he might be blackballed come initiation time unless he crawled to Tim and told everything he knew. The note said that Jim was being watched and his actions were being noted, and he should take the "honorable path."

Oh, yeah, more "secret oath" stuff.

Jim came to me two days later. I had him make an appointment with me in my room later that evening. When he arrived the room was arranged so that he would sit in a chair looking up at me as I sat on my desk. The only light in the room was a bare light bulb stationed directly behind my head so that Tim had to look up at my dark shadowy form and see light shining all around me, while he sat in relative darkness. Pink Floyd's creepy ANIMALS album was playing the song "Dogs." It was at the place in the song that said, "And as you lose control you'll reap the harvest you have sown. And as the fear grows the bad blood slows and turns to stone, etc."

Voice cracking, Jim spilled his guts in one long run-on sentence. He mentioned the "guys," the glass (he was the culprit, of course), bitter remorse, and was visibly trembling. His eyes were even a little watery when he told me that he didn't want to get blackballed. Naturally, I assured him that everything would be fine. I threw on the lights, patted his back, told him that he owed me a beer, and that was that.

Well, Tim M., it's not the personal and private version of hell week I was expecting after all that build-up, but I must say I heartily approve. Negativity only begets negativity. It's more important we have truth coming to the light of day, forgiveness and a softening of hearts. Something I said in my previous book bears emphasis: Too many people involved in "survivalism" are also caught up in "vengeance trips." But a vengeance trip is usually *counter* to sustaining your life force, which is the goal of all forms of survivalism.

It is important to say this because there are SO MANY EASY WAYS TO FUCK SHIT UP with stuff you find in the dumpster. I would urge people to be visionary in their goals. The World Trade Organization, for example, deserves whatever it gets. The fellow whose shopping cart rolled into your car? Tell him to buy you a beer, just like Tim did.

Why, according to myth there was a goddess who wanted to kill off the whole human race by blood sacrifice, but the human race offered up some 6,000 jars of beer dyed the color of blood, and she changed her mind.

Thank goodness for all of US, huh? In the spirit of that goddess, I say when somebody has offended you and you want to shed blood, tell the offender that he *owes you a beer.* Or accept his offer *to buy you a beer.* Or suggest you both discuss the matter *over beer.*

Not TOO much beer, of course.

It would be bad to, for example, *take his wife*

hostage OR *point an M1 rifle at his chest* OR *tell the police he is making drugs in his apartment* OR *vow to destroy his reputation forever.*

Bob? Jim? Are you both hearing me?

(Sorry, reader. An *internal* matter among notorious cult authors.)

Let's continue with Tim's nice letter, shall we?

Ahh, the trash, faithful provider of all that is necessary to obtain justice, solve mysteries, and provide the common man with the means to take on uncommon foes.

I wonder how information diving in Washington D.C. would be? Did you read about the SOLDIER OF FORTUNE personnel who "info dived" a BATF office? He landed office floor plans, code names, informant's addresses and a whole bunch of great stuff, which he promptly turned over to a lawyer, who gave it to the proper authorities, who in turn chewed the asses of the BATF for casually discarding such sensitive intel, and the BATF, in turn, probably planned raids on the homes of all FORTUNE subscribers. (See James L. Pate, "SOF Exposes ATF's Warbirds," SOLDIER OF FORTUNE, October 1995, p. 80.)

A great example of effective "info diving," Tim. And the kind of media manipulation I was talking about. I give this soldier of fortune a twenty-one can salute.

BOOOM—BOOOM—BOOOOM!!!!

You would think that truly sensitive garbage would always be disposed of properly, but that's just not the case. What causes such security breaches is THIS attitude: "Yuck, who would go in the GARBAGE?"

Revulsion is relative, and all a result of social programming. There are individuals who become faint at the process for making "head cheese," a dish which includes the facial meat of a pig, and yet they smack their lips at the thought of delicious hog butt (i.e., pork chops). They are *horrified* at the thought of eating insect-based delicacies such as crispy locusts, chocolate-covered ants, mealy worm cookies (please, Mr. Chef, second helpings?), and yet they enthusiastically spread bee vomit (i.e., honey) upon their toast in the morning. They discard perfectly good vegetables which GREW IN DIRT ENRICHED WITH COW SHIT, FOR GOD'S SAKE, because the veggies are a little less perfect than plastic fruit.

Only a deliberate act of self-blinding keeps us from seeing that, all the divelong day, everybody wallows in shit. The water you drink, the food you eat, I don't care if it's Zam Zam water given to us by angels and only found in obscure sections of Middle Eastern food stores. I don't care if it's haute cuisine served up at $200 a plate. All of us roll around in filth like swine. We eat shit and drink piss. That is our lot as carbon-based life-forms. (Pynchon suggests making the transition to silicon-based life-forms, and I think he makes a good point.)

When it's all over, you are food for the worms. The soul, if there is one, might be your only chance for real purity. *Mock the dumpster diver with your Madison Avenue-based values?* Nay, remove the log of feces from thine own eye.

When I see a thing, I strive to see its characteristics and not shadows of my own subjectivity and social programming. So it is that one perceives "shining opportunity for a serious security breach" instead of "yucky stuff I doubt very much anybody would pilfer." So it is one sees "rope I can use to reach one of the suitcases from my crashed aircraft" instead of merely "mint-flavored dental floss." In a survival situation, somebody with this objectivity, which comes from a deep connection with the life force and clear consciousness, looks around and sees "weapon, weapon, weapon" instead of "chair, candlestick, violent intruder breaking down the door, trapped and helpless, helpless, *helpless!*"

This planet's life-sustaining oxygen started as an excreted waste product of certain simple life-forms which, at that time, dominated the planet as we now dominate it. But you do not go around thinking, "Piss and shit! Ayeeeeee! My lungs are a toilet!"

At an atomic level, the distinctions between food and shit, Coke and Pepsi, right and wrong . . . disappear. Cannibalism, dumpster diving, keeping kosher—it is all relative. The ability to eat almost anything is one of the primary characteristics of life-forms that just don't go extinct. Figuring out how to exploit a resource that is *everywhere* that hardly anybody else is exploiting or values is the path of a dominant life-form who will one day be as rich as Madonna.

Arise, Trash Hopper! Rise above it all, pure and white like the lotus, uncontaminated by the muck from which we draw our sustenance. In this day and age, to eat food from a dumpster and avoid paying another dollar plus tax to the big and evil Global Biz Gov is better than fasting for Lent, superior to keeping kosher, more practical than fish on Friday, and your soul will rise for your efforts, and the Dumpster Goddess will clutch you to her sweet milky bosom.

Ponder this, if you will. The human race has the brain and outlook of a hunter-gatherer even though, relatively recently, many of us have become urban dwellers deriving our life force from ethereal abstractions (being paid to program computers, for example). Why did a book about dumpster diving sweep the world and become an enduring "cult classic?"

Because it completes the circle! The dumpster diver is an urban hunter-gatherer. Even those who would never sully their hands upon the lid of a dumpster have read the words of a dumpster-diving guru and thought, yes, this makes sense. This is how life flows. This is how you find goodies and avoid bigger, meaner predators. And what was really important all along? Passing on my genes and caring for the next generation by obtaining and sustaining as much life force as I can. Through my descendants, my memory will not fade from the earth, my desires and consciousness will continue to live, and my progeny (mental, physical and spiritual) will pray and *think* my soul up through the heavenly spheres.

Where was I? Tim!

As to the old Bible pages, John, I recently dived some very hard-to-get university letterhead and envelopes. Perhaps a "Dear Librarian, it has come to our attention . . ." letter.

If you want to push your luck, Tim, go for it.

Finding blank letterhead used to be a great opportunity, and it's still convenient for really great pranks. Merely finding letterhead makes people start thinking of great pranks they never otherwise would have pulled. But now, with computers widely available that can replicate most anything to an amazing and eye-fooling degree, one doesn't even *need* the blank stuff. You just need an example of a particular letterhead or form and some computer skills. So, while blank letterhead and forms have become LESS valuable to divers in the past seven years, the value of ALL letterhead and forms has actually increased substantially.

We must no longer be mere divers.

But cyberdivers

I hope this letter finds you well, and a hearty congratulations on your new baby boy, Alexander James!!! We have two of our own. Meagan is 20 months and George is 5 months. Meagan is being potty trained and "laid an egg" on the kitchen floor while I was writing this letter.

Child rearing, parenthood, new mommies . . . boy, I could go on and on about THOSE subjects.

You found a wedding ring for your wife Tina? That's too cool. I occasionally find jewelry in old discarded purses, after garage sales, or when people clean house before a move. I'm keeping my eyes open for jilted lovers throwing away boxes of expensive gifts but haven't had the luck.

I'll have to tell the story of the wedding ring here, Tim. I was diving in El Paso and found an old garment bag with an elaborate hand-stitched pattern. Though it was pretty worn, I thought it might be worth a few bucks. It was full of odds and ends, too, so I took it to my apartment, only yards away, to let my wife look through the girlie stuff and see what she might find.

I should point out that, up to this point, my wife wore a nice-looking but completely fake wedding ring purchased for $30 at the Fort Bliss military commissary. It turned her finger army green, but it was the best I could do at the time. My word, what my dear wife went through those first days of our marriage! I used to liberate Polish sausages and oranges from the army mess hall at Fort Sam Houston and smuggle them to our little home. To this day I have a penchant for "cargo pants" with big pockets useful for smuggling. I think it would be a good idea for somebody to make cargo pants out of the fine cloth that goes into, say, an Armani suit. You just can't have too many pockets, even at a formal dinner party, as a statesman making a speech, or as a defense lawyer arguing before the court.

Sigh. Our first married home together was a

converted toolshed somebody had the audacity to rent as an "apartment" to the spouses of soldiers, but it was a hell of a deal. My wife was fond of the free samples at a local grocery store, which could be cobbled together into a whole meal, and she would gather fallen pecans from the trees in the neighborhood and crack them open, watching the television I dumpster dived when I was a senior at Concordia College.

Experts estimate that American casualties could be as high as 50,000 in the first 48 hours of a ground war. Rear areas may not be safe from SCUD missiles, and while a full-scale attack on the United States is not possible, terrorist attacks are a very real possibility . . . .

(Later, I would donate that television to a subversive cable access show in Seattle called DEFACE THE NATION, to become part of a cluster of salvaged televisions broadcasting live during the show while the host commented acidly upon the "corporate mass media." I will discuss the principle of "maximum subversive impact" of donated dumpster goodies when I explain how I mailed the Lesbian Avengers my left ball.)

We had the world's cheapest wedding, Tina and me, as you might expect of the master diver. The only witness was a Spanish-speaking janitor who wandered in during the ceremony in the church parlor. (In Texas, a wedding can be legally witnessed by the minister performing it.) We took what little money we had, rode the bus into downtown San Antonio, and sat near the Riverwalk. We lacked money to do anything but ride home again. But, like my father, I always have a scroungy trick up my sleeve, and nothing motivates me like the need to provide for my family. I saw some money in a wishing well, and—when nobody was looking—I retrieved enough coins to buy bean burritos on special at a nearby Mexican restaurant.

Only months later our luck was on an uptick, and we had a wonderful apartment facing the Franklin Mountains, a fridge full of food, appliances galore, and my wife was taking college classes. She was looking through that lovely old garment bag and found a pile of junk jewelry down at the bottom. One of the pieces of junk was a nice gold chain. (Readers will recall my words, "Thar's gold in them thar dumpsters!!!") And right in the middle of the pile of junk, tangled up in the gold chain, was a gold wedding band with five little diamonds.

A diamond wedding ring.

From the dumpster.

I guess those coins from the fountain still had the power to grant some wishes.

Pentium computers . . . diamond rings . . . information to shake the foundations of society . . . Hollywood movie scripts worth a small fortune . . . gold-plated Oscars . . . *that's* what dumpster divers are finding in the trash.

Let the prissy doubters laugh their asses off. The universe has its chaotic ups and downs, and those who are in tune with the life force . . . who perceive its true nature, its ebbs and flows . . . who know how to use a rolled-up newspaper as an ersatz funnel to put gas in a stalled car, whether that car is a Mercedes or a Pontiac . . . when the chips are down and life itself is at stake, it is the *weirdos* who always have a life-sustaining trick up their sleeve. And it is the picky eaters and overly choosy specialists heavily invested in the status quo who will ultimately belly flop into extinction, and *that* painfully.

The ability to capture free-floating life energy by opportunistically exploiting waste is literally one of the oldest and best tricks on the planet.

What is your future, Trash Hopper?

What does your dumpster-dived fortune cookie say?

"You will conquer the universe, you will paint the galaxy green with Dyson Spheres, you will become and remain the dominant life-form, FOREVER, sitting with your dive stick at the right hand of God."

CHAPTER 8

Special Guest Chapter

Industrial Diving

Mark from Texas is such a fan that, when I told him about my antiquated computer, he sent me a Pentium computer with a Deskjet printer that he had scavenged, programming and setting up everything for me so that all I had to do was plug it in. This guy is such a master that he gets his own chapter, and truthfully it's probably the most valuable chapter in the whole book.

He even gets the straight-up letters instead of the italics!

While I write books with "political science fiction" ideas that will, I believe, change the universe, and I use dumpster diving to supplement my lifestyle, often invoking the lessons of my dumpster-diving parents, this diver from Texas took a few of the master diver's teachings–promptly threw about half of them away–and colonized a whole quadrant of the refuse universe all by himself. I made him send pictures just to PROVE what he was saying because it was so incredible.

The man does $5,000-a-night routes and gives his friends Pentium computers for gifts.

And here Mark is spilling the beans because, despite the fact his secrets are making him a fortune, he still cares more about the planet and the future of the human race, and for that reason (plus the money I negotiated from Paladin for his guest chapter), he's willing to spill the beans.

So without further ado, here's Mark the Big-Hearted Texan!!!!!

(Sound of the class members banging dive sticks on dumpsters.)

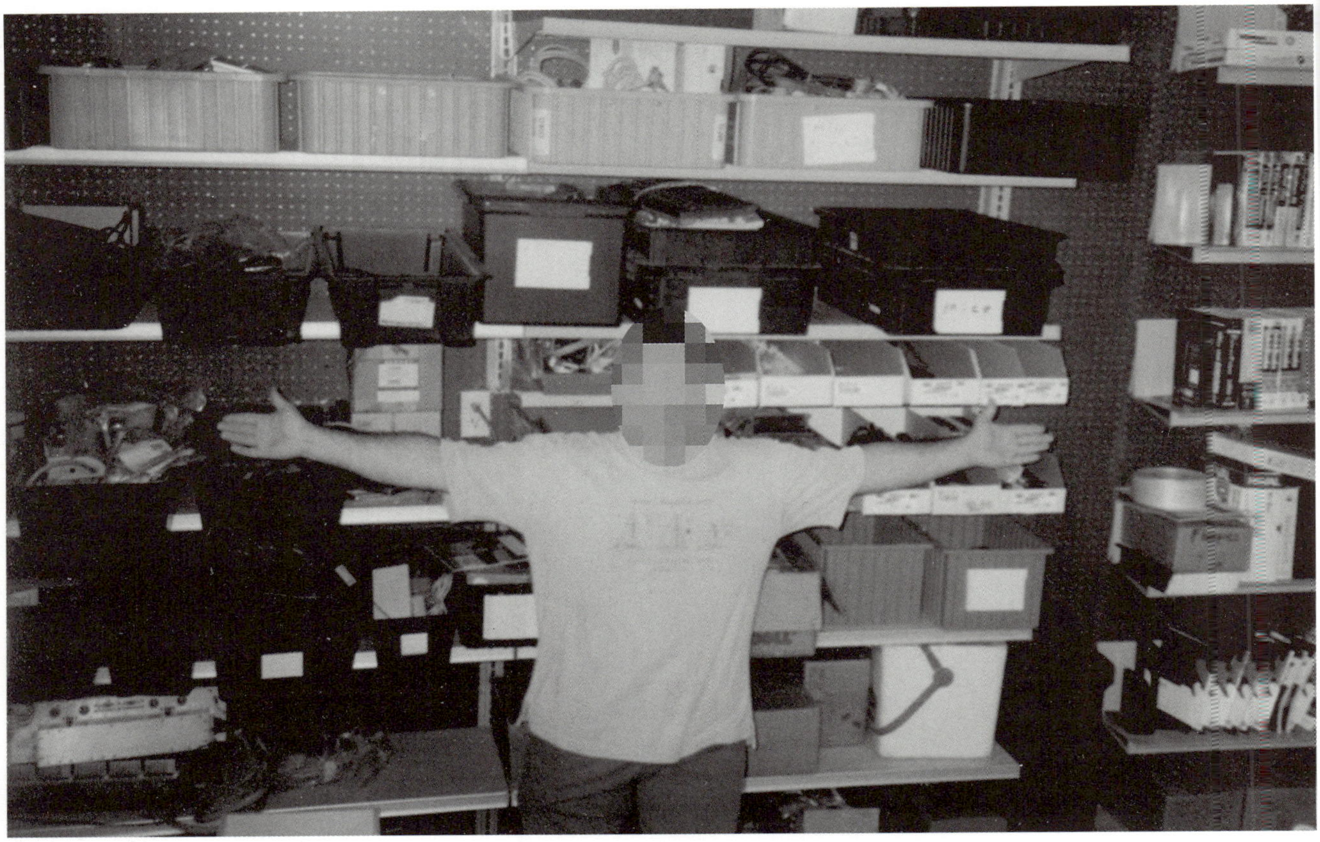

My favorite wall. The shelves behind me hold more than $15,000 worth of dumpster-dived goodies. By the way, the building where these pics were taken is my new workshop. It cost $6,000 to build, and yes, it was paid for entirely with "trash money." One more thing: all of the metal shelves and wall brackets were dived too.

I am a dumpster diver. I make my living by rummaging through the garbage of others. While I don't use this as an icebreaker at cocktail parties, it is a fact nonetheless. I am a soldier in the War on Waste, a believer in the Religion of Refuse, a priest in the Temple of Trash. I am a humble student of the Master John Hoffman. Yet, as it is with all students, I have left the Master to follow my own road and thereby I have become my own Master. I walk a road "separate but equal." In a world of paths less taken, I choose the least of all.

Industrial trash.

I have to admit that I have removed a large part of the Art from my repertoire and refined the Science. While this may have adversely affected the spontaneous, fun nature of my profession, it has been a serious boon to my bottom line. I stick to places that most dumpster divers avoid like the plague, spending all of my precious diving time in industrial districts and warehouse back lots. Most people think these places are only full of cardboard, used motor oil, and broken old pallets. Oh, the irony is just delicious!

I'm making a fortune.

I started my dumpster-diving career as an after-school hobby. In high school, porno is gold. Those magazines might as well have been neatly stacked piles of $5 bills. It's good to know some things never change. I would also acquire cool clothing for school and Hostess snack cakes.

I gradually evolved from residential to commercial trash, but I stopped there. After high school my diving tapered off until I did very little at all. I followed a more traditional pattern and got a regular job, then married and had a couple of kids, now up to four. I always thought of DD though, and often wondered if there were perhaps a way to make a living at it.

Fan mail? Nope. These are the envelopes in which I received payment for the items I sold on eBay. This is one month's worth of envelopes.

A LAPSED DIVER FINDS HIS STEELY BALLS

My wife and I started diving at the local landfill, and the path to enlightenment began. You see, the landfill left all the trash from Friday and Saturday unburied until Monday morning. So, naturally, we would dive on Sunday.

Mark, it's not breaking the Sabbath to do the Lord's work.

One fateful Sunday, we came upon a large pile of new industrial bearings. These were the really big suckers, the 10 and 12 inch dudes, and they were still in their boxes. It turned out that the local Union Carbide plant had decided to scrap all of the "old" unused bearings.

Mark, why does it not surprise me to find out that UNION CARBIDE did something evil? Can you say "Bhopal," boys and girls? I KNEW YOU COULD!

It ALSO just so happened that we lived in oil field country, and we ALSO had an industrial supply house in town to boot.

Well, needless to say we hauled those bearings out of there until it got too dark to continue. Mind you, the landfill gates where we had to park were about half of a mile from the actual dump site, and believe me, it was all uphill. I went back out at 5 AM the next morning and kept hauling out bearings until just before eight. As I pulled away from the gates, sure enough here came the landfill guys. I had an escape window of about ten seconds . . . they weren't too hip on the idea of our scavenging.

We sold the bearings for more than I was making in a month at my regular job. And that's when my idea was born. Leave the rat race and become a specialist at *industrial* diving . . . .

I do almost all my work (and that's what it is—

a job) on an industrial scale. I do not dumpster dive as a hobby or sideline. It is all that I do. I tell people that I run my own business and I do "industrial recycling." Euphemisms are occasionally necessary in our line of work, aren't they? I make my living from warehouses, wholesalers, and some retail. By focusing on what I consider the "cream" of the refuse crop, I eliminate the less productive "nickel and diming" of the standard residential and commercial style.

Don't get me wrong! A good living can be obtained, and lots of fun can be had, from diving only apartment dumpsters, but one could never consistently bring in the level of merchandise and money that is available through industrial diving. On a standard residential/commercial diving route, one could expect to earn between $75 and $200 a day depending on the particulars of the route. When I dive my route, I earn between $200 and $2,000 per day, every time.

Here I must, ever so gently, take issue with Mark. I mostly agree with what he is saying, but his later writing reveals that Mark is both focused upon and very familiar with industrial processes. Somebody in a residential diving area might make a fortune by focusing on some other particulars like, um, well, small-scale suburban blackmail, to choose a bad example. Also, as Mark continues, let's never forget that Mark lives in the Texas Silicon Valley, where there is also a lot of petroleum. Can Mark's methods and success be replicated elsewhere? I believe the answer is yes, but that remains to be seen. As for myself, the first day of law school is a month away.

I have not earned less than $300 on a single route in over four months, and my average is just over $600 per run. The most that I have earned from a single run is $5,300. Not bad for three and a half hours work! By the way, my route consists of thirty stops and is in a constant state of fluctuation. I drop old places and add new ones as disposal habits change. Seasonal patterns affect my route as well. Never become complacent and stop looking for new additions to your route.

I work by a set of rules. First, I'm going to introduce you to the TOP TEN PLACES TO GO INDUSTRIAL DIVING:

#10 Wholesale clubs Sam's
#9 Food service vendors
#8 Shipping/moving companies
#7 Toner/ink jet cartridge vendors
#6 Government warehouses
#5 Computer R&D companies
#4 Office supply houses
#3 Electronic wholesalers
#2 Computer retail outlets and upgrade shops

And my absolute favorite place to dive is . . .

#1 Telecommunication supply wholesalers/retailers

You have undoubtedly noticed that I am rather partial to electronics companies. Well, what can I say? We're living in an electrical world, and I'm an electrical kind of guy. I believe someday our children will be able to dumpster dive from home through the magic of virtual reality. The image the feel, the smell, the profitability, and the very *excitement* of DD itself captured for all to experience in the comfort of their homes, safe from all the dangers and diseases of yesteryear.

Sob, sob. Sorry, I get a little choked up when I talk about this stuff. All right you got me, I tend to just go where the *money* is . . . .

And the money is in electronics.

WHOLESALE CLUBS

While this type of location is actually a retail/wholesale hybrid, it is worth the time and any trouble that you might encounter. What you can expect to find will vary from location to location, depending on the type of merchandise that the store handles. But the number one thing will, unfortunately, be PACKAGING MATERIALS, so unless you really NEED moving boxes, plastic wrap, Styrofoam peanuts and the like, just ignore most of this stuff.

Um, Mark, I actually met a guy who dives and SELLS that stuff, especially those stupid Styrofoam peanuts that ought to be against the law.

We Americans are a little anal when it comes to the packaging of our stuff. Great news for cardboard companies but terrible news for the

Where's the fire? These are fire putty sticks, and they are used in network installations when a hole is drilled in a fire wall. Retail is $10 per stick, but your cost is free!

environment. But what you are looking for is this: any item that was used as a "display item," which is an item that they took out of the box, wrote off as "store use," then put on a shelf for the customers to look at. These might include a new printer, a lamp, VCR, calculator, vacuum, toaster, a 10-disc CD car radio, or even a brand new computer. You won't know until you start looking, and the items and quantity will always vary depending on the season.

If you want to make it easy to watch for items, simply go into the store and ask the department managers when they are expecting to get new shipments of their seasonally shifting merchandise. They will simply think that you are an interested customer and will often tell you the exact day that they will be getting the new stuff. Sure enough, that very same night, odds are they will be disposing of the old stuff, right into your arms. THANK YOU, MISTER __________ !

Here, Mark uses the name of a prominent chain store owner who, I think, is inclined to be less than thankful. Might I also add that, in my experience, these waste patterns ain't so simple and predictable, but Mark has done things on a bigger scale and in a more specialized, focused way, so I'm willing to believe his experience. I'd also hesitate to talk to a store manager, but Mark does these things in a bigger town and, well, he's not on television all the time, tagged as a famous dumpster diver, so he can get away with it!

Some of these stores are a pretty good source of food as well. I no longer have grocery-only locations on my route, but you can bet your ass I won't pass up food if I find it. Other things that you are looking for include overstocked items and "RTV." RTV means it is to be Returned to Vendor—in other words, to the company the store purchased it from. (Watch and you will see it written on a lot of items.) Many times the item is

NOT being returned because it is damaged but because it was on display or was opened by a curious customer, though that's NOT what the retailer said when they returned it to the vendor, who placed it in the dumpster, where you are even now diving.

It has been my experience that the vast majority of things thrown away by wholesale and industrial locations work fine. In fact, I am shocked and dismayed when I test an item and, occasionally, find it nonfunctioning. I have to ask myself, "Now why the hell did they throw this piece of shit in with all this good stuff?"

You're right, Mark. They should MARK IT somehow to save you effort. They might write, for example, BROKEN followed by SMASH THE WTO.

FOOD SERVICE VENDORS

While I do not frequent grocery stores any longer, I do regularly dive these little gems. These are the companies that deliver food, supplies, and equipment to restaurants. You can count on getting boxes of food just like the stuff from your favorite eating establishments. Yum, yum! It usually comes precooked (restaurants just love precooked food; it saves them time and the effort of having to be actual cooks) in quantities to feed fifty or more. Steaks, shrimp, pasta, exotics (dumpster-dived calamari, anyone?), and just about anything else you want. Most of it is strictly first-class stuff, yummy. Also expect to find lots (and LOTS) of plastic utensils, napkins, place mats, linens, table centerpieces, and everything else you see in restaurants.

Mark, sounds like this really blows the magic of the dining experience. All for the better, I might add.

I once found a new four-burner Bun coffee machine with only minimum shipping damage. Rest assured that the small diner that bought it from me for $300 was not the least bit concerned with the scratch on the side of the unit. I also recently acquired all of the SOS pads and 3M kitchen scrubbers that I will need for the rest of my life.

Mark, I think I know somebody who would trade you some GUM!

SHIPPING/MOVING COMPANIES

Wow! I can't say enough about these places. Talk about a surprise everyday. It's almost as good as Christmas.

If you ship something and the shipper then tells you it got lost, it didn't. They damaged it and then threw it away, because on insurance reports, "lost" looks better than "damaged." Insurance people make me ill. These places are great for high-dollar items, especially electronics.

TONER/INK JET CARTRIDGE VENDORS

OK, here's a weird one. I recycle used, yes used printer/fax/copy machine cartridges. The people who buy these items from me clean the cartridges, then refill them with ink or toner and then sell them back to the public for half the price of a new cartridge.

I like these people a lot. They are helping the environment AND the economy. Not an easy thing to do at the same time. Used cartridges command between fifty cents and $20 each depending on their type. I like inkjet cartridges the best because of their size. (You can fit fifty into a bread box.) Finding these things is as hard as falling down. They're everywhere you look. I mean, who doesn't have a printer, a fax machine, or a copier?

Read this next part closely to understand it. You may have to read it twice because what Mark has done is so clever.

The best places to acquire these are actually the companies that either sell them or remanufacture them. The people who sell them seldom recycle them, but they do like their buyers to THINK that they recycle them. They take them in under false pretenses and then throw them away. The people who remanufacture them, on the other hand, usually only remanufacture certain ones, BUT they like their customers to think they do them all, so they throw away the ones they don't use. I scoop them up and then sell them to another company who wants THAT kind. While there, I get the ones THEY don't want and sell them back to the OTHER company. It's kind of

Black gold! Used inkjet cartridges fetch good money. What you see here on the table brings in about $350.

like robbing Peter to pay Paul, only in this case, Peter and Paul are both paying me.

I like it this way much, much better.

The Master is BREATHLESS at the brilliance of his pupil!

GOVERNMENT WAREHOUSES

Do I have to explain this one? HELLO, it's the government! All they do is waste things: time, money, and the blood of young men! Each warehouse will yield its own strange mixture of stuff. The Land Department might have a whole bunch of desks and one box of porno mags. The Department of Health might have computers, lab equipment, and a bunch of old shoes. I don't have the slightest idea where this stuff comes from and, honestly, I don't think THEY do either.

Most of these are easy targets with little or no security, BUT a few will have guards or fences or both. The regional IRS warehouse is in my dive city and has two forty-yard dumpsters that are filled and emptied every other day. Great stuff! Only one problem: armed guards. Yes, that's right, your tax dollars at work—two men with guns protecting IRS garbage from what exactly? And I'm not talking about the files or papers. I'm talking about the old desks and chairs and shelving units.

Other than the occasional weird place, these are pretty tame and pan out well.

I would like to add to Mark's words that at the time I received his letter I was actually a government official myself. Hard to believe, but I was the first Green Party member elected in the entire state of North Dakota. I served on the city council of Grand Forks, North Dakota, where I fought against the new landfill and pleaded for more recycling. Winning with 29 percent of the vote, I failed to beat a recall effort with 44 percent of the vote. While in office I didn't sacrifice the blood of any young men. Of course I wasn't in office very long, and all the young people are leaving Appalachia on the

Here are a couple of items I found together at the same time. Both are brand new, and both are used in high-end networking applications. The one in my left hand is worth $300, and the one in my right hand is worth $600. By the way, I have four more like the one in my right hand.

Plains in droves for better wages, better climate, and a better culture.

I might suggest this as a reformist measure to reduce government waste. If it is just too much trouble to dispose of certain items as government surplus, then set them out near the unguarded refuse areas of the particular government office or warehouse near a sign that reads as follows:

SUB-SURPLUS GOVERNMENT ITEM FREE FOR SALVAGE AND RESALE. PLEASE TAKE AND USE AT YOUR OWN RISK.

I might suggest the passage of a federal law that enlightened divers might wave in the face of government security guards and landfill officials, guns and all.

"I claim this property in your [here insert "landfill" or "very guarded dumpster"] under the sub-salvage section of the federal code, 611-97. I hereby warn you that to interfere with my salvage is to commit a federal crime. Could you help me with this lid, officer?"

COMPUTER R&D COMPANIES

Nice place; lots of computer and electronic parts and equipment. Varies depending on the location, but you can expect to get computer boards, integrated circuits, fiber optic parts and equipment, phones, and anything else that requires electricity. Sometimes they even dispose of cutting-edge technology.

OFFICE SUPPLY HOUSES

These people kill me. They will throw away a package of fifty new pens because one is missing. The term "mark it down" is not in their vocabulary. Here you can expect to find paper, pens, files, folders, post-it notes by the box, electric pencil sharpeners—just about anything that's office related. You can be assured I don't buy office supplies. I do, however, sell them by the box, and a large box that

I found all of these (15) software kits on the same night at the same place. They sell for $500 each. Damn our "disposable" society!

Fifty-seven Microsoft NT Workstation CDs. Same story: one night, one location.

What's this? A brand new Sony wireless microphone receiver that costs $900. Whoa! This little baby was sent to the wrong person, so they returned it to the warehouse. As we all know, once it goes back to the warehouse it's considered to be junk, so they threw it away! Right into my arms

is. You can also find bunches of used inkjet cartridges. These places are sure bets every time you go, but watch for seasonal swings that make them even better targets, like around the time school starts. You better bring a friend . . . and a truck!

ELECTRONICS WHOLESALERS

Computers, video games, VCRs, TVs, home and car audio/video equipment are what you can expect to find in these places. It is unfortunate, but you can also expect to find compactors in most electronic sales locations. But there are ways around these artifices and contrivances of evil. John touched on this in his first book, and I liked the idea so much I adopted it.

Um, Mark? That book, as well as this one, is for education purposes only. Don't be, like, going out and applying stuff you read.

My take is this: if the compactor becomes uneconomical or a nuisance, then they will do away with it. The best approach is one of stealth. Do not make it apparent that you are tampering with the unit. No super glue in the buttons, and don't cut the power cord—this is dangerous! Do not give yourself away or they will use every means at their disposal to stop you or even catch you on surveillance and prosecute you "to the fullest extent of the law."

The best way to attack these Satanic mechanical devil dogs is to bring them down from their weakest point—their hydraulics system. Since the hydraulics are usually the first thing to fail anyway, this is seldom seen as the assault that it actually is, just a normal malfunction. After consecutive malfunctions, the company will often decide to do away with the compactor and go back to locking the dumpsters. AND WE ALL

These are *new* Microsoft Operating System programs (Windows 2000 servers). Please notice the other boxes behind me. There was a total of 1,040 units! That came to a grand total of $100,400. Not bad for thirty minutes work.

KNOW HOW EFFECTIVE THOSE LOCKS ARE, DON'T WE FELLOW DIVERS?

Loud sounds of laughter from the class, hoots and hollering. . . .

"OK class, simmer down! Let's all listen to Mark, here."

To sabotage the hydraulic system, you need only a good, strong needle. A leather sewing needle works best, in my opinion. Find the hydraulic hose (it's the messy one) and give it a couple of good jabs near a fitting. Once the fluid has drained out, the compactor will be inoperable and the company will have no choice but to call in a dumpster for temporary use until the compactor is fixed. The trick to this is to make sure that once the mechanic has fixed the hose, you IMMEDIATELY attack it again. It only takes a few of these attacks to really piss off a company and have them up in arms.

Vary your attack point. Don't use the same spot on the hose every time. Play around with different hoses and fittings. Throw in a good "chaffing" problem every once in a while, too. That's when hoses or wires rub together and damage themselves. It usually takes some time for this to occur, but the natural processes can be speeded up with a small metal file. Make sure that you clean up the filings from the hoses; don't leave them behind as evidence.

Once you have achieved your goal, then you can really get yourself the *best buy!*

(Loud coughing from the Master.)

Typos, Mark. You almost capitalized something corporate by mistake. Besides, I would consider "best buy" to be more of a discount retail store than an electronics wholesaler per se. But as long as something with brand name recognition is painfully twisted into a joke about dumpster diving, I'll let you get by.

These are Lucent wire-wrap blocks that are used in network applications with telephone company equipment. I find these by the case. They sell for about $10 each.

COMPUTER RETAIL OUTLETS AND UPGRADE SHOPS

Same as above—computers, computers, computers. Hey, other things are fun to dive, but the real money is in electronics. Most computer outlets and upgrade shops do NOT use compactors. The ones that are quite new will use compactors, but you know what to do about THAT. You will find yourself running into competition, but the answer is simply to beat them to the goodies.

And now, my number one diving spot . . . .

TELECOMMUNICATION SUPPLY WHOLESALERS/RETAILERS

Oh, where to begin? No one dives there, and nobody bothers you while you dive. These people throw away more great stuff than you can believe. Fiber optic parts, microwave relay equipment, analog transceivers, 19-inch equipment racks, conduit, ducting, wiring, cables, connectors, electronic monitors, and high-end electronics of every kind. If the box has a *scratch* on it, they will throw it away. One of the places that I go fills up a forty-yard dumpster every third day. Talk about a lot of money in the trash. WOW!

Seek these places out and start to learn about them. If you think these types of items are over your head, you're wrong. You can learn this stuff in your sleep. Keep the catalogs you find in these same dumpsters and you will have a ready-built reference to use. If you need an answer to a specific question, just call the 800 number in the catalog and they will tell you everything you want to know about the item. After all, they want you as a customer of their products.

Dream on.

The *really* smart people don't pay.

A WORD ABOUT RETAIL STORES

A retail store manager would rather pass

Telecommunications gear is great. This is a Fujitsu rack mount chassis—new. It brings about $450 on eBay.

More telecommunications gear. This is an AT&T rack mount connection panel used by the phone company, worth $250.

shards of broken glass out of his tight, puckered anus than mark down a returned, cosmetically damaged, or out-of-date item, wouldn't he? So in turn, he orders his minions to destroy all usable items prior to their being discarded.

Thank God for lazy stock boys who, when not being watched, simply throw the stuff out instead of taking the extra time to make sure that the precious trash is pounded to smithereens. Or perhaps they just have a CONSCIENCE and can't bring themselves to sledge hammer a brand new 50X CD-ROM that's worth $65 and is still in the box. For whatever reason, from time to time items slip through the web of destruction that is the shipping/receiving area of a retail store's back room. God Bless America.

I am sad to say that I have found items that have been so badly destroyed that it makes me wonder about the guy who smashed it up. Is he mentally ill? Is he ex-mafia? I swear, if this stuff had been a human being instead of a VCR they wouldn't have been able to identify him with dental records.

I'm not trying to discourage you from retail stores, only to warn you about what you will find. Sad but true.

MARK'S DIVING PHILOSOPHY

Sure, you can make money by dumpster diving standard residential, commercial style, but why do that when you can make two to ten times more by simply changing your target areas? The diving doesn't change, only the locations and profit.

To drive home this point, I want to tell you about a guy I once knew who did the flea market thing. He would go out and buy items from garage sales and auctions, then go through all the time and trouble of cleaning, repairing, and testing them before he would sell them at his booth a couple of weeks later.

One Saturday, I casually asked him how much he made for the day and he went into such a long and drawn out oration that his own CPA (if he'd had one) would have found it to be rife with details. But I did a little figuring in my head and came up with $3 an hour. When I mentioned this to him, he said that sounded about right. I told him that he would be better off working at McDonald's for minimum wage. At least he could afford to feed his family better than he was doing at the time.

Needless to say, this pissed him off royally and he never spoke to me again. The truth hurts. Yet the fact remains, if you are doing your DD in a disorganized and haphazard manner, then you are pissing your effort into the wind.

Following my path is not for everyone. Many people who DD love the antiestablishment feel to it. I mean, what is more in-yer-face than digging in somebody's trash? Some of these people feel that using DD as a business makes it more "mainstream," more capitalistic.

But, hey, I help the world every day by recovering hundreds of pounds of resources that would otherwise be buried in a sanitary landfill and then reintroducing these items back into the economy, thereby creating income not only for myself but also adding a boost to the rest of the economy. I should wear tights and a cape with a little logo upon it, superhero style. I like what I do and I'm damned proud of it!

I should add that Mark has contributed to "in your face activism" with his gift of a computer and printer to the Master, and I wouldn't doubt that when Mark encounters activists or causes that he particularly supports, once his own family is provided for he might throw a computer or box of office supplies this way or that. Mark talks all profit-oriented and survivalist . . . and then, when the chips are down, he's one of the most big-hearted and generous people you're likely to meet.

I start my run at 9 PM. I drive for 30 minutes to get to "the city," but man, is it worth it. My little town just does not have enough industry to support my diving needs. I have spent the last two and a half years tweaking this route to get it where it is today, but I am always flexible and ready to change. You never know when you might stumble upon a gold mine, and it's easier to realize when you have your eyes open.

When it works out, I take one of my three sons with me each night. We rotate from the first born on down to the last.

Mark, I treat my three-year-old to little "peeks" in dumpsters. "Anything good in there?" I ask. And if he

A glimpse into one of my four storage units. This unit is 10 feet by 30 feet and filled to the top, as are all of my units. Supply is never an issue. Note the large spool of communications cable at the bottom right of the photo and the box of toner cartridges to the left of the cable.

identifies something good or just something he's curious about, daddy goes after the item. Once I poked at a bag and said, "Alex, that bag has got MONEY inside of it!" And Alex was amazed as I pulled out some coins inside a drinking glass, coins for him to put inside his little denim pocket. Now he is under the impression that many dumpsters, if not most of them, have cash for the taking. Ah, this is true quality time with the little guy! This, and driving around North Dakota looking at abandoned houses that can be purchased for a few hundreds bucks, land and all . . . .

By the way, my kids are EXCELLENT help while diving. They are learning the ropes from the bottom of the dumpster up. Kids are great. They ask lots of questions and retain things like a sponge. They make me feel like Master Po in "Kung Fu."

I find things interesting and things weird. Recently, I found an eight-yard dumpster filled to the rim with computer chip fabrication lab coats. They were separated by size (S/M/L/XL) and were not only freshly laundered but also folded and sealed in plastic bags to keep them lint free. These things are made to be nonconductive, thus no static electricity, no sparks, and no fried computer chips. New, these lab coats cost $90 each. I found over 200 of them! Now, I will only get $10 to $12 each for them but, hey, you do the math.

My pride and joy find of all time has to be when I found eight NEW, IN THE BOX ParkerVision CCD video teleconferencing wireless remote, full pan/tilt/zoom cameras. I sold them for $600 each in four days. This happens to be the same run when I totaled $5,350. I also found a perfect-condition working Pentium laptop computer that night. Gotta love them computer stores!

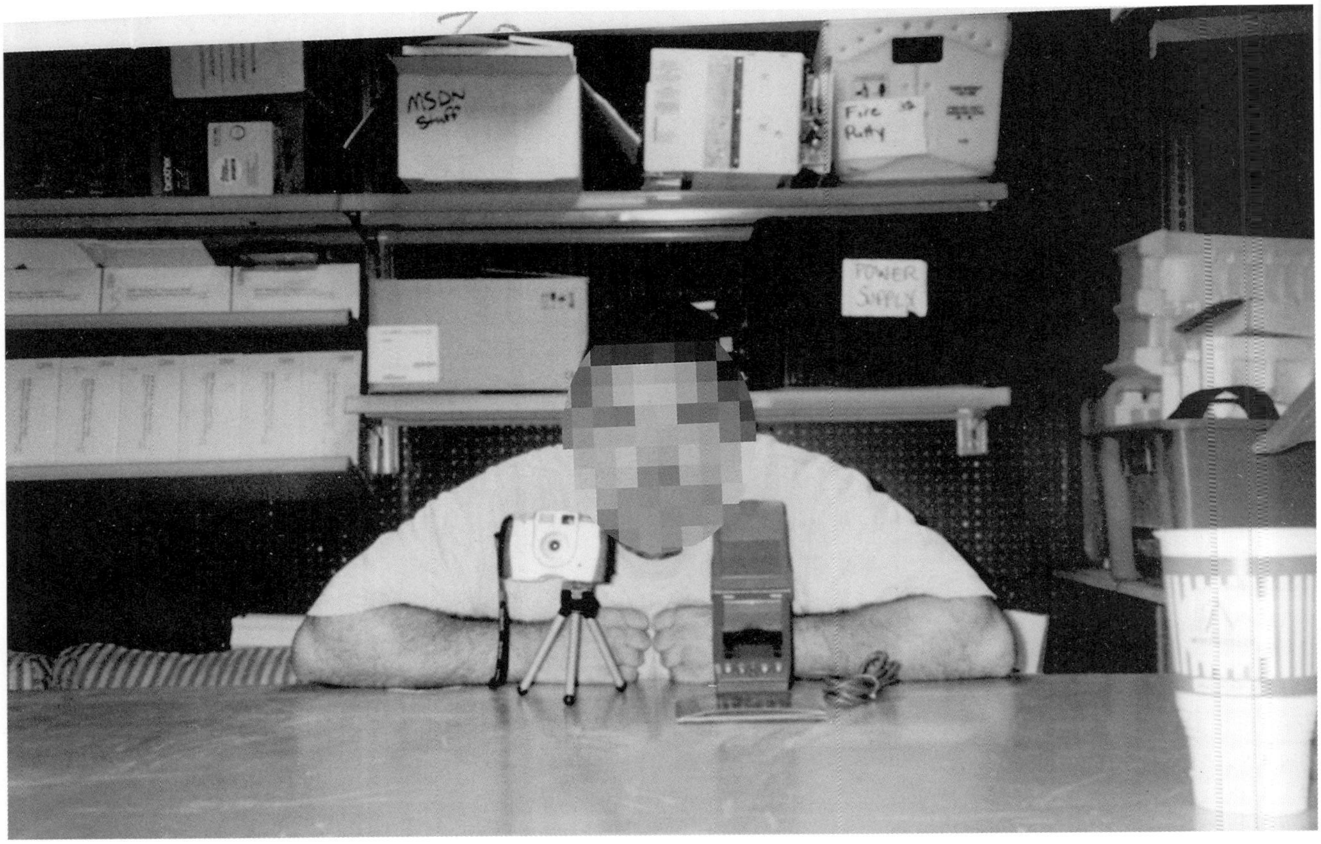

Herein lies a funny story. I found these in the same dumpster on different days. I am *positive* that the same guy at this electronics lab threw them away. On the first night, I found an empty Palm IV box, but next to it was this guy's working Palm III. Out with the old and in with the new. The next night I found a box for a new digital camera. Yep, next to it was his old digital camera. Some people are so short-sighted that I find it hard to understand how they are able to walk around without falling down all the time.

And Mark brings one of these fine offerings of pure Pentium to his Master! God, somebody find me a discarded supermodel . . . .

The cameras came from a rather unexpected place. I picked them up from a shipping company. And I actually found out the *story* behind these items, and I'll tell you how in a moment. But here's what happened.

It turns out the shipping company sent a truck full of electronics to a company on the east coast. Being the thorough company that they were, the receiver made sure to go through each part of the shipment prior to signing for it. Low and behold, some of the items had been damaged in shipping. Well, this would not do for the receiving company, and they refused to take possession of THE ENTIRE SHIPMENT. The poor trucker had to load all of the pallets back on his truck, drive them all the way back to Texas to the shipping company warehouse, and unload the items . . . where they sat in limbo for four months.

Well, after four months the insurance company agreed to pay for the damaged goods and THE ENTIRE SHIPMENT WAS THROWN IN THE DUMPSTER, SURPRISE, SURPRISE. This is the way of shipping and insurance companies. HELLO, STUPID, THE *OTHER* STUFF IS FINE!!

I know that I shouldn't hate these guys, but they are smug and arrogant and stupid to boot. Never buy stock in their companies. In case you are wondering how I have all this information, it's simple: I called and asked.

Hold on. I know what you're thinking, but it *really* does work like that. I simply called ParkerVision and inquired about the cameras and their value and asked if maybe they would like to buy them BACK since they were NEW and STILL IN THEIR BOXES AND EVERYTHING. They weren't interested, but they were very free with their information and the sordid history of the cameras while in the possession of the shipping company. It seems that they had also been contacted by the insurance company about the prospect of a buy-back on the cameras. The insurance company got the same answer as me, and what does an insurance company need with $10,000 in cameras, right?

Um, Mark, it could look up its very large collective ass.

So of course, the insurance company called the warehouse and told them to "dispose" of those troublesome and costly cameras for them, thanks.

I hate these people but love them, too. It's a real duality problem for me. On the one hand, they keep me in business; on the other hand, their stupidity pisses me off.

It sounds kind of like being a counselor and taking care of the mentally ill and substance abusers. Only in this case, the mentally ill are passing themselves off as quite sane while abusing the whole planet.

In the first book, John mentions recycling aluminum cans and copper wiring to make a little money for gas. I sort my aluminum into five categories and my copper into three, but I also recycle brass, stainless steel, titanium, and high-end alloys (magnesium, tungsten, nickel, and the like). But my favorite is GOLD.

Yes, GOLD. The old prospector in the cartoon in John's first book was right. THAR'S GOLD IN THEM THAR DUMPSTERS!!!

Here's the secret. Green computer boards and CPU chips have gold in them. The amount varies by unit, but the money is good. Finding old computers and computer components is about as hard as falling off a log. With the advent of new technology, the old stuff is discarded. I strip down the old PCs, 286s, 386s, and even some 486 computers to scrap them out for gold content.

HOWEVER, NEVER STRIP DOWN OLD, OLD APPLE/MAC OR PRE-IBM STUFF!!! There is a specialty market for "vintage" computers. The stuff from the 1970s is pulling in big bucks. Remember the old suitcase-size portable computers? Many of those are selling for more than a new Pentium III laptop. So don't tear ass through any old, old stuff without checking or you'll be kicking yourself.

Here is the basic, current scale for the value of scrap computer boards and CPUs:

Motherboards	$1.00 per pound.
PC cards	$2.00 per pound.
Backplanes	$2.50 per pound
Misc. boards	$.50 to $.85 per pound.
CPUs	$55.00 a pound

YES, THAT'S 55 DOLLARS A POUND!

Take into consideration the fact that your average computer has about two and a half pounds worth of boards in it (AND A CPU!) and that you can EASILY average ten to thirty computers per month and you can soon see the motivation behind getting these things. Bear in mind that the older the computer, then the bigger the boards, and thus the more they are worth.

I personally salvage around 600 pounds of boards every three months, plus the CPU. That's an extra $1,200 check for about six hours worth of work. My metals recycling nets me another $200 per month. It used to be twice that, but the metals markets have been weak for the last year or so.

Back to warehouses for a while. It might enlighten and amuse you to learn my eight rules of industrial diving, contrasted against the six rules of warehouse people.

1. NEVER assume that because it is in the trash it has little or no value.
2. NEVER assume that the same people who threw it away won't buy it right back.
3. LEARN the value of everything you can. Knowledge is power. Save every industry catalog you find in the trash. Keep them organized and on file.
4. BUILD a network of buyers and stay in regular contact with them. If their buying needs change, be aware of it.
5. Be FLEXIBLE. You must bend like a reed in the wind, Trash Hopper!
6. Get ORGANIZED or die trying.
7. DIVERSIFICATION is the key to ultimate success.

8. Get a computer and GET ON THE INTERNET.

These rules work for industrial diving partially because of the rules that warehouse people work by, which are, as near as I can tell, as follows:

1. If it hasn't moved in three months—throw it out!
2. If it has no paperwork with it—throw it out!
3. If I lost the paperwork for it—cover mistake by throwing it out!
4. If tired of looking at it—throw it out!
5. If it will cause me to do some work in the future, or if I suspect it might—throw, throw, throw it out!
6. If it's brand new, costs a bundle, and it's still in the box—THROW IT THE HELL OUT!

Before giving you my final salute-ation, I will give you a little insight about these rules. AS GOD IS MY WITNESS, I can't even count the number of times I have sold items back to the same company that threw them away in the first place. The key is this: don't try to sell them back to the guys in the warehouse! Instead, go to the main office, which is in a different building or even a different town.

Ask for the person in charge of acquisition and tell him or her what you have, and mention you just KNOW they use this kind of item anyway, so why not skip the middleman and buy it more ECONOMICALLY directly from YOU? You don't need to do this in person; the telephone works quite nicely. Nine times out of ten, if you are a good salesperson they will buy it there and then. Soon it goes back to the warehouse where you got it in the first place.

No kidding. I swear. I've sold some companies the same item three or four times because the warehouse guys kept throwing it away!

Mark, what really cracks me up is that what you are describing, while it sounds very fraudulent, is probably PERFECTLY LEGAL.

A simple but overlooked rule is to ALWAYS remove the stickers, labels, or other identifying marks from the item or packaging PRIOR to selling it or they'll think you're in cahoots with the warehouse guys and you'll find yourself in a fix with a lot of explaining to do. This will increase the security at the warehouse, which will piss off the warehouse guys. Bad idea. Remove the identifying marks.

I make it a priority to know something about everything. My goal is to be able to walk up to every object in the universe and know what its function is and have at least a rudimentary idea of its value. Again, I can't say enough good things about catalogs. SAVE THEM ALL AND SORT THEM OUT!!!

One of the cool benefits of catalogs is that you gain insight into the buying habits of the company that threw them away. You will have a pretty good idea what items they buy, sell, and dispose of and it's nice to have a price list for the items that you dive in those locations.

THE DUMPSTER THAT GOT AWAY

I have to confess something here and now . . . how I screwed myself out of a $50,000 dive. They say confession is good for the soul. Well, I sure hope so.

Oh, boy. Here goes.

Bless me, Master, for I have sinned.

There is an electronics product out there called an IC, which stands for integrated circuit. In a nutshell, ICs are the little computer chips that you find in damn near everything made today. Hell, even some TOASTERS have them.

Mark, I'm still waiting for a toaster that will toast just one side of the bread so I can make a buttery little toast boat by pressing in the hot, doughy side. I had a half-functional toaster like that at Caisson Hill and it was heavenly.

Anyway, I find ICs from time to time, new and in their "tubes," but I was always unable to sell them at all. I finally gave up and began to ignore them altogether. Complacency breeds disaster, however. Not too long ago I found a dumpster LOADED TO THE TOP with new IC chips in their little plastic tubes. I decided to give them another try, so I grabbed only a small box full and headed off to the next location on my well-worn route.

Later that week, I did a little digging on the Internet and found a company out in California

These are the "Quick Start Guides" that come with the software for a new computer. Most people think these are worthless. They are not! I sell them for $3 each to my software buyer. It might not sound like much, but when you find 1,000 at a time, it sounds a lot better.

This set of shelves is for software and computer books. I commonly find the covers removed from the books, but 90 percent of the time the cover is lying in the dumpster right next to the book. What, you never heard of tape?

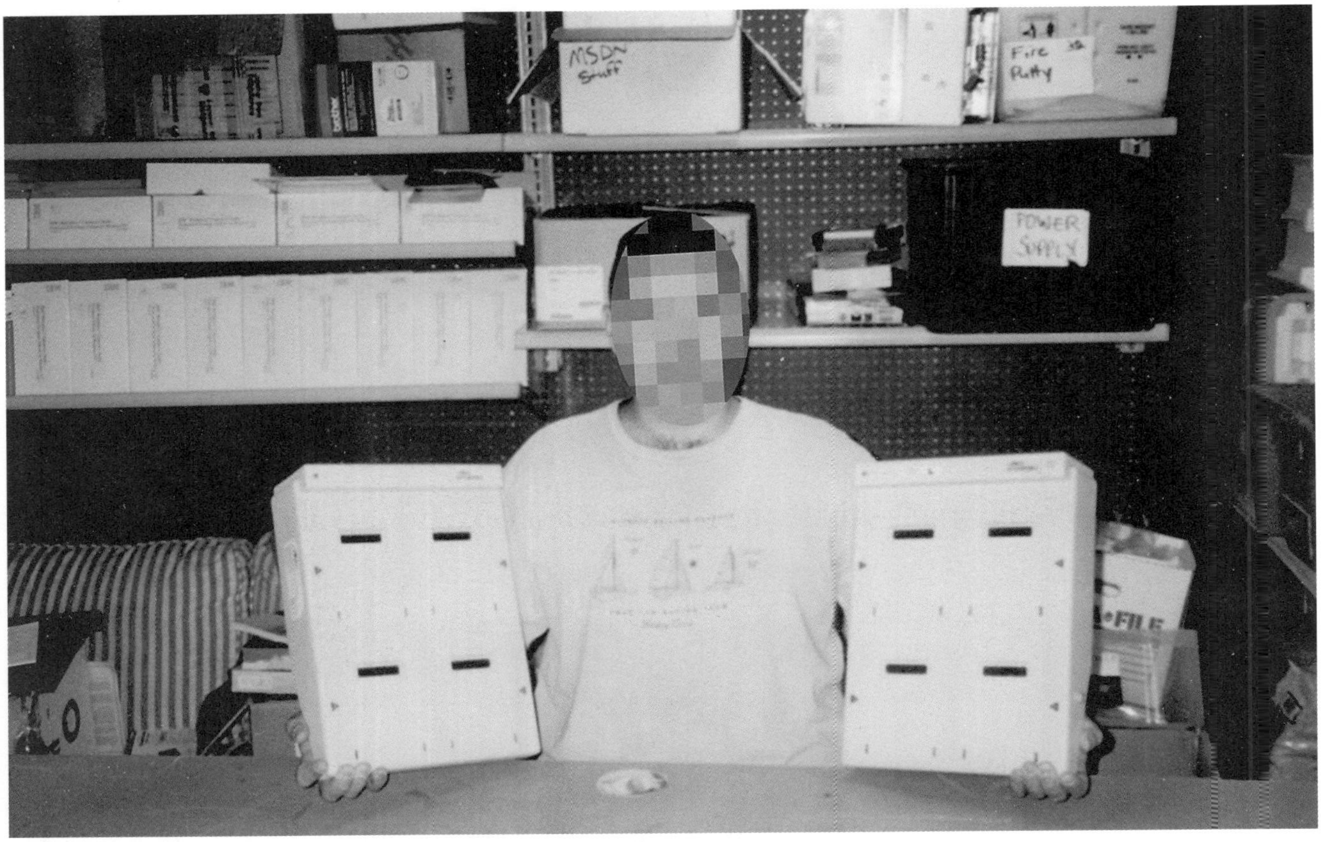

Here is a pair of IC (integrated circuit) "chip" programmers. They do not have the IC trays with them, and that is why they were tossed out. Even without the trays, they are still worth about $100 each on eBay.

that might be interested in the ICs. I contacted the company and not only were they interested in them but they Fed Ex'ed me a check overnight. I mentioned the other chips to see what their response might be. Hold on to your asses: THEY WERE WILLING TO PAY ME $50,000 FOR THE ICs THAT I HAD LEFT BEHIND!

I ran to my truck at that very moment and went back to the dumpster, BUT IT WAS TOO LATE.

Ayeeeeeeee! The horror, Mark, the horror. I'm sorry. I'm trying to keep it together. THANK GOD THIS IS A BURDEN FOR YOUR SOUL AND NOT MINE! Oh, you'll be working off THAT karma for a long time, Mark!!!

The precious little IC chips were already forty feet underground in the local landfill. I've been pretty much kicking my own ass nonstop since then. Let my mistake be your teacher and remember this lesson.

Remember that if you can't find a buyer for industrial goodies that you dive, that only means you haven't looked in the right place just yet. The streets really are paved with gold, at least for those who have the vision to see trash as the treasure it actually is.

Here, here, Master Mark. Now, I want to bring up the possibility that a market might be found for some goods in Mexico. Seeing how Mexicans cross the border JUST TO SCAVENGE DUMPSTERS in border cities like El Paso, from that I extrapolate that the discards of certain American industrial processes might be valuable to (can I say it without giving offense?) less-developed Third World industry, and Third World dollars, once converted from pesos or whatever, spend just as well as American green. The Master will chuckle when the day comes that a Mexican diver successfully asserts NAFTA provisions to get his goods across the border. A few months ago I scrounged some ornamental bricks from a construction site in Winnipeg, and that was pretty cool . . . bringing dumpster-dived goods through U.S. Customs and declaring them as such. I got through with no problems.

"The New Byzantine," © 1994 Edward Doherty II, constructed from computer circuits and motherboards.

I just want to throw this in here. My old army buddy Edward Doherty II made a crucifix out of old motherboards, and he was offered $50,000 for the piece—entitled "The New Byzantine"—by an ultramodern Catholic Church located near a campus. The deal fell through at the last minute. However, my buddy Eddie later observed Mexican artisans making these same crucifixes out of motherboards and selling them. Eddie swears he was the avatar of this concept, and knowing the dates involved—this crucifix piece was first shown in 1994—I believe him.

Which goes to prove a point I want to make right now. Beyond the ease of the good money Mark from Texas makes with industrial discards, there is an even greater form of ease, and that is making money from art using the same materials. Mark has all but rejected the art part, and I must use this moment to balance that point of view. The art part matters. The art is good. But Mark is making a major contribution upon the Science end of the Art and Science of Dumpster Diving. In the sequel to this sequel, should there be one and should it be entitled PEOPLE OF THE DUMPSTER, *perhaps we will be hearing from an artist or two.*

Just as I was going to send this to you, John, I made a new find that I have to include. Remember the eight expensive cameras that I found in a shipping company's dumpster? Well, I found twelve more the other night. These are a different make and model, but they're worth even more. As I type this I am making preparations to deliver them in the morning to "the city." Upon delivery, I will collect $9,000. Not bad for a day's wages, even for an old trash hound. Take care out there, HEROs!

Our special guest chapter writer Master Mark the Big-Hearted Texan informs me that HERO stands for Help Earth, Recycle Often. The Master bows before the student, Mark from Texas, who has become The Master . . . .

CHAPTER 9

Children of the Dumpster

The first chapters of this book were written on an old Tandy 1000 SL. This last chapter is being written upon a Pentium, a gift from Master Mark, with letter style choices like Papyrus, Poor Richard, and Castellar. There is not a doubt in my mind these words will one day be read upon planets that circle distant stars, by children who can trace their lineage to the Master Diver.

And perhaps you will be their ancestor as well, our DNA mingled as our thoughts, in a sense, become mingled as you read these words written upon angel light in Algerian typeface. What an amazing, wonderful universe we live and dive in.

LEAVING LAS SEATTLE

Moving again, lightening up my possessions by selling them, giving somebody a good deal, letting go of sentimentality (thank god for pictures), hauling odds and ends in those last moments to secondhand stores that support Youth Leagues and Chicken Soup Brigades, but more often applying my Theory of Maximum Subversive Impact by giving as much as possible to Heather, Heather my Hero, Heather the Lesbian Avenger.

I met Heather on December 2, 1999, my 34th birthday. The day before I had been shot by a rubber bullet and gassed twice with unknown chemical agents in the streets of Seattle. It was a relief, actually, being shot and gassed. It took my mind off the fact my wife had left me on December 1, catching a flight out of a city that was still smoldering, fleeing for Grand Forks, North Dakota, and, well, conventionality amid the prairie grass and the fucking sugar beets, taking my precious son, Alexander James, to a place I call Appalachia on the Plains.

To be shot and gassed and see 14-year-olds throwing themselves in front of armored personnel carriers was soothing and distracting like a morphine dream.

Sometimes I could get above and beyond my personal pain for as long as 45 seconds because I could find PERSPECTIVE in the pain and struggle all around me. The chaos of the anti-WTO uprising seemed like a manifestation of the chaos inside, a cry of understanding and comfort from the universe itself. The riotous world of the anti-WTO protests made perfect sense, subjectively.

After all, if the world inside me has gone crazy, why should everything around me seem normal? Why should traffic flow and shopping continue? If Icarus falls from the sky into the ocean, why should ships blissfully sail, heedless? So it is that I teach you media manipulation . . . let the outer world reflect your inner world. If you are mad at your boss, let the world be furious with him as well and mirror your inner fury.

The most terrible crime of our consumer culture is this: we are forced to inhabit an outer world which is nothing like our inner world yet which constantly presses upon us with its shrieking consumerist demands, and ultimately it presses upon us so hard that we are swallowed up in it, our existence negated. Ask anybody what they dream of doing or becoming, and so often IT ISN'T THIS THING THEY ARE DOING RIGHT NOW.

For almost everything we are doing, using, and consuming there is a price that presses upon our time, our energy, our very existence. What happened to the splendid tree, from which we could just pick the fruit and eat? Why should we labor by the sweat of our brow and bring children into the world in pain? Why should we not enjoy eternal bliss and comfort, here and now, not in some future utopia promised to us by the preachers and the monks in saffron robes?

This is the wonder of dumpster diving, my friend. The tree is still growing. It is pushing up through the cracks in the sidewalk and offering us its fruit.

Free.

That wonderful word, free free free. Take and eat; this is the fruit which has been given unto you, and it contains the seeds which will grow a new world, a paradise, an outer world which will reflect our inner world, and our inner world will be one of pure bliss.

• • • • •

On December 2, the radicals who weren't locked up in the King County jail mixed freely with the more orthodox pro-labor elements, and at a pivotal moment these radical elements broke free and ran through the streets, bravely trying YET AGAIN to get near the convention center to shut down the hated World Trade Organization, encountering police lines but flowing freely like water, redeploying with the speed of words transmitted by cell phone, but the truth was . . . it was spontaneous, not a brilliant act of redeployment using modern technology, no matter what the police say in their after-action reports.

It was mad, made up in that moment, and some people knew where they were going but most didn't, and everybody was shouting and chanting, and some would sit down in front of the police lines, and some wanted to go the other way, and at a certain moment one element got around the police lines and seized the courtyard of the jail, yet again, and my smaller element saw the damned Westin Hotel right in front of us, unguarded, and it was like, hey, ISN'T THAT THE WESTIN WHERE LOTS OF THE DELEGATES ARE STAYING?! UNGUARDED? JEEZ LET'S FUCKING GRAB IT!!!

We swarmed up the front steps and took positions, and the police line didn't seem to be forming very fast around us, but I figured it would. Who knew if more people and supplies would be able to get through the police line? The most dedicated troops were locking their heads to the door of the Westin with titanium bicycle locks, brave child-warriors fighting for the sea turtles. I looked for somebody who appeared to be experienced in these matters, somebody with "command presence" . . . and there she was.

Almost fashion model beautiful, wearing a black beret and her departed mother's scarf, Heather the Lesbian Avenger sat upon the stone steps in cool repose amid teenagers swirling like packs of squirrels, quietly surveying the police line, her eyes utterly calm and calculating.

I knew that lovely face. I had photographed her the evening of December 1, looking on as the trade unionists threw fake steel girders into Puget Sound. At that time she was nearly naked from the waist up, her pierced nipples not quite covered with electrical tape, a militant warrior woman, one of the goddesses of the Battle of Seattle. Upon her chest she had written a defiant challenge to the police: GAS THIS!

Our position was tenuous. The anti-WTO troops had an average age of, like, 16 or 17 and, at that moment, could be scraped off the front of the Westin like a barely clinging barnacle unless we FORTIFIED OUR POSITION. It didn't seem at all ironic to think of Drill Sergeant Dozier at Fort Jackson, who taught me how to create fortified fighting positions and had once said, "You can always improve your situation, and you should, you must. You can dig grenade sumps, put in communication lines, booby trap the perimeter. ALWAYS, ALWAYS THINK OF HOW YOU CAN IMPROVE YOUR POSITION, TO FIGHT BETTER, TO SURVIVE AND, YES, TO BE COMFORTABLE AND KEEP UP YOUR MORALE."

"What do you need here?" I shouted to Heather, above the din, above the mind-altering omnipresent drumming and chanting of, "THIS IS WHAT DEMOCRACY LOOKS LIKE!"

"Water!" she whispered, and there was a moment of desperation in those strong, flinty eyes of hers.

Of course, I thought. We've been running through the damned streets for half an hour and now we need something to drink. Water!

Cops near our perimeter were radioing unknown, menacing things. I dragooned David Todd O. and told him to come across the street with me, to Bartell Drugs. I figured we might be able to get one, exactly ONE load of supplies through the police line before it hardened up. To everything there is a season, turn turn turn, and this was not the season to dumpster dive but to whip out my Platinum Discover Card and purchase gallons of water and many boxes of oatmeal cookies. Water, sugar, and carbohydrates. Let the goddess in the black beret be pleased with my humble offering. We hauled ass past the police and passed out supplies, using Heather as a kind of central distribution point.

"OK, now what do you need?" I asked Heather, as the young troops drank the water and dived into the calories, their faces almost too beautiful to look upon.

"You got any smokes?" she asked.

Smoking killed my father right in front of me. But it sure gave him something to do with his hands during that dreadful island-hopping campaign and has to be seen in that context.

"I'll see what I can do," I answered.

For our troops, I did something I've never done before in my life. I panhandled.

There were crowds of onlookers, many with sympathetic faces, and I systematically hit all of them up for cigarettes, quickly adding, "Candy, gum, anything you got to support the protesters who are locked to the door, ready to be arrested." The cold hard universe, in its mercy, kicked down packs of smokes. And, incredibly, a WTO "gift box" with smoked salmon from one of the delegates!

And STREET MAPS! We could have used a few of those earlier in the demonstration, but oh well! The troops lit up, which caused a schism and forced the civil disobedience "happening" to divide into smoking and non-smoking sections.

"Is there a bathroom around here somewhere?" Heather asked, as I made yet another delivery of goodies.

Oh, god. Now THAT was a problem.

I went into an alley where I knew there was a restaurant and scored two white plastic five-gallon buckets. Then, in rough but enthusiastic Spanish, I convinced a group to donate their red banner, which we would put right up front, slogan facing outward. So, stringing the red Hispanic banner between two potted shrubs to provide a measure of privacy, we established a crude bathroom upon the very doorstep of what I believe was a four-star hotel. Leaflets and items of clothing were ripped into toilet paper.

We now had a bathroom. THAT made us a civilization.

I made a foray of a few city blocks and walked into one of the nicer restaurants in the city, where glasses and plates clinked musically and well-dressed yuppies were discussing yuppie things, including those DREADFUL demonstrations, oblivious to what was happening only a few blocks away. Walking into a rest room that featured piped-in video of soothing ocean scenes, I liberated bathroom tissue and, on the way out, fine linen napkins, white bread, and pats of butter.

I will never forget the sight of Heather gratefully pressing a roll of toilet paper to her gorgeous face. The table linens were made into bandannas as the threat of a gas attack suddenly grew. More shirts were ripped up and pressed into service as bandannas. The media were everywhere. Later I would read in the local papers about the makeshift toilet upon the steps of the Westin and think to myself, THAT'S MY REVOLUTIONARY POTTY, BABY!!!!

The police threat ebbed and flowed, but the cold, hard steps were sucking the life and energy out of those brave kids. There was a lot of talk about giving up the Westin position and consolidating with the other protest at the King County Jail. I didn't want to voice my objections too strenuously, but I knew that, tactically, we were wasting a lot of effort occupying the courtyard of the jail, weeping and wailing over our prisoners while the World Trade Organization went on about its business. The best thing we could do for our prisoners was to keep trying to shut down the WTO, not camp in front of the damned jailhouse singing "Kumbaya." The fate of the Westin occupation depended upon keeping our morale high at that moment.

I went into a nearby alley and I began dumpster diving. For boxes. The ancient excuse—when you're trying to pretend you're not really up to anything, you're just looking for some BOXES. I hauled boxes until every protester had a piece of cardboard between his or her body and the cold stone steps. Incredibly, morale began creeping upward and there was less talk of giving up the position.

And we held. We fucking HELD that position, and some elements of the Teamsters showed up—after I cornered them and prevailed upon them and *convinced* them as a union shop steward to help out these kids in front of the Westin—they showed up with hot coffee, peanut butter and jelly, and bread. And I was making sandwiches like a motherfucker, PBJ for revolution.

"Who are you with?" asked a young man with a golden star in the middle of his forehead and a kind of, well, call it pizzazz. "Are you with the Direct Action Network?"

"I'm just an old dumpster diver and an army medic," I said.

"But who are you WITH?" he insisted. "You're so good! You must be with somebody!"

So I replied, "Let's just say . . . I'm with LOGISTICS."

"Logistics?" he asked.

"The art of moving supplies to the front, where they are needed," I explained. "People come to protests and what do they bring? Words, words, words. Sometimes what's really needed is *stuff.* That's what *logistics* is about. Get the *stuff* the hell to the front to support the fight."

He gave me a big kiss on the cheek.

"You're the best LOGISTICS ever!" he gushed.

And Heather was watching, grinning, licking jelly off her fingers.

Pieces of cardboard from a dumpster, to keep the precious heat in your body at the moment of struggle, when you need to be brave. A white plastic bucket to relieve yourself. Some linen pressed into service as a bandanna. A bite of *smoked salmon,* a moment of pure luxury amid the squalor of revolution. Peanut butter and jelly. How much the little things we take for granted MATTER when the fate of the whole world literally hangs in the balance.

If you watch the famous video about the five days that shook the WTO, some of the very last scenes take place upon the steps of the Westin, the position we seized and that I struggled to fortify and hold. I was there as the words that jolted humanity were uttered, a message read from a pager by lame duck King County Councilmember Brian "The Derd" Derdowski.

"The World Trade Organization talks have COLLAPSED!"

Wild cheers begin to erupt around him, but still there was MORE to relay to the crowd, more good news, unbelievable news that would rock the world.

"There will be NO . . . NEW . . . ROUND!"

Then those brave children who had been shot at point-blank range, gassed with unknown agents, possibly NERVE GAS in the Pike's Place Market, were dancing, dancing in giddy circles and chanting a new chant.

"We changed the world! We changed the world!"

• • • • •

MY THEORY OF MAXIMUM SUBVERSIVE IMPACT is simple. Little things mean a lot, and there are so few radical activists devoting their lives to struggling against the powers that be, so few people DOING SOMETHING instead of putting out more words, words, words.

Dumpster diving is DOING SOMETHING. Each of the things you find has the potential to DO SOMETHING. Each of these little pieces matter and has world-changing potential, chaotic butterfly power.

Giving a radical some laundry soap, a couple of CDs, a decent pair of boots is an act that resonates for eternity. For example, if I give the Salvation Army an old roll of electrical tape, they will probably sell it for 25 cents and, well, that's 25 cents for the Salvation Army, hurray. But if I give that roll of electrical tape to Heather, you can bet she will use it to barely cover up her nipples so she won't get arrested for flashing a television camera. Add up all the moments individual television viewers will spend gazing upon those breasts and it will add up to lifetimes expended in a moment. A lifetime well spent, I can personally assure you. Each of the individual human minds—like iron filings aligning themselves toward a magnetic point—will perceive the same image and incorporate it into their individual essence. Without a word to express something, or an image to which one might point in explanation, how can meaning even *exist*? Thus the roll of tape RESONATES in the mass consciousness. It achieves MAXIMUM SUBVERSIVE IMPACT, becoming as valuable and important as it is possible for that roll of tape to be. History books will attempt to explain how the world came to be as it has become, and your hoary hand will point to the words on the page: *at that moment, I was there. I handed her the very roll of tape!*

And the children, gathered around to hear the stories of the family for their education and edification, will think, "Our family is connected to things which matter, as evidenced by our connection to this important moment in a book of history. Why, without the tape, how could that moment have even happened? My relative has done great things, and I will do things which are greater still!"

To a lesser degree, if I give Heather laundry soap to get the chemical agents out of her sweet girlie underthings, if I give her calories to propel herself into the street, where she hands out lollipops in Pioneer Square and raises dyke awareness, those items I have bestowed upon her have a great degree of subversive impact, more so than if I gave them to my innocuous and normal, friendly neighbor Fred. Finding the MAXIMUM impact is not easy, and frequently one must settle for an impact that is merely good.

But since I want my actions, even my little actions that make up so much of mundane existence, to MATTER and to RESONATE, I think for a moment about where each little piece should fit, what its maximum potential might be.

Moderates are plentiful, but radicals are precious and few. When one radical smashes the window of a bank, a hundred moderates rush in and try to explain, to apologize and justify in the same breath. It takes radicals to MAKE SOMETHING HAPPEN. So it was that I gave Heather the old wooden headboard that lived in the closet, the one missing a ball.

"I think I packed the ball with my stuff that went to North Dakota," I explained. "I'll find it and I'll mail it to you in time for your fund-raising rummage sale. What wonderful jokes you will be able to make!"

Truthfully, I was a little afraid inviting Heather into my apartment for the first time. She was, after all, a Lesbian Avenger and I was a guy whose wife had just left him for, we must assume, some good reasons. Part of me wondered if Heather would cut my throat and take my credit cards. Instead, she sat on my couch, drinking beer, watching old tapes of Tina and me promoting the dumpster book together in Hollywood.

"Is this painful for you?" Heather asked, her voice tender and full of gentle concern.

"It is very painful," I admitted.

How strange that, in our relationship, I played the very traditional role of PROVIDER as well as the role of FELLOW RADICAL. But it was precisely what I needed at that moment, somebody to tell me, not in so many words but in her actions, the following: "You are a wonderful provider and a dedicated radical, both qualities I respect. And I enjoy being with you, even saying flirty and outrageous things to you, despite the fact I am an avowed lesbian. I feel safe and comfortable alone in your apartment with the lights off, on your couch, drinking a beer."

When I imagine the wonderful future of the human race, I imagine that an offshoot of the Lesbian Avengers will have their own planet, where they will reproduce with parthenogenesis, and it will be a beautiful and well-managed planet where heterosexual males will feel safe and comfortable and will want to spend their tourist dollars, though GOD HELP THEM if they should get it into their head to rape one of the locals. So it is that I use the Lesbian Avengers as a prime example for my theory of Maximum Subversive Impact and suggest that, when giving away energy and resources, you find a group with equal potential to change the universe, and not settle for—gag!—MODERATES.

Later, I would write Heather a poem called BEGIN MOVING TOWARD LOS ANGELES, and part of it goes like this:

All revolutionaries instinctively raise up a goddess figure
The mother gives us permission. She will be proud and clutch
Us to her milky breast as we swing from a tree, spit upon
Called a traitor
Though she could be any young woman in the crowd
Who likes to show off, but at that moment she is the very
Goddess Herself in the CS gas, red hot pepper gel, the rubbery
Bazooka gum bullets bounce right off her as she raises her
Bloody fist, defiant, and as we huddle in People's Park
The Denny Regrade, the Sunset Strip . . . any place can be
A battlefield, especially the Red River Valley
Any place can be so important
Babies are conceived in cabbage patches
But not in Heather.

• • • • •

Little things mean a lot. Anything can be a lever to move the world, even a roll of electrical tape or an old wooden headboard missing a ball. The question is not whether you have what you need to change the world, only if you have the vision.

In my previous book, I mentioned that misguided people—especially teenagers—sometimes set dumpsters ablaze. During the battle of Seattle, dumpsters blazed up, and at one point a protester said, "That's kind of dumb, isn't it? I mean all that paper could be *recycled!*" And, while videotape rolled, the other protester said, "You've got to understand. This will be seen all over the world."

In other words, rather than being made into mundane forms or toilet paper, the paper in that dumpster (actually a recycle bin) will be converted to light and heat, and the paper will convey a message that will still be heard in ten thousand years. There is a time to salvage blue glass bottles from the recycle bin and make a lovely and fragile stained glass window, and there is a time to *ruthlessly break windows.* A time to recycle glass bottles and a time to hurl them. To everything there is a season. What better symbol of defiance than a blazing dumpster?

A FRIGHTENING THOUGHT THAT DOESN'T FIT ANYWHERE

Every day, the bacteria that make us sick are getting tougher as they overcome the newer, more potent antibiotics that we are developing. I've heard it said that we should be judicial about taking antibiotics or, for example, dumping antibiotics into cattle feed.

But what about when we flush old antibiotics down the toilet? Could it be that this common practice is one of the ways the bacteria are gaining an advantage? Isn't the sewage full of bacteria? And aren't we giving them a laboratory to develop resistance every time we flush old pills and capsules? To a lesser degree, tossing old antibiotics in the garbage could be creating the same effect.

It is, once again, this unwillingness to deal with our waste products . . . to even THINK about the forbidden, nasty, dirty subject . . . which makes our civilization unable to perceive what seems, to me, rather important and obvious: that we should immediately cease the practice of flushing old antibiotics down the toilet, that THIS could be the primary breeding grounds of superbacteria.

I have to wonder what planetary breakthroughs we might make if we would, as a civilization, become "potty trained." Right now, much of our civilization is like an infant, crapping out its waste into a giant plastic thingy, to be disposed of by mommy and daddy and not thought about anymore, certainly not something for which we are RESPONSIBLE. Dumpster divers and those who recycle are like those few cells in the brain of the infant that have an idea, a notion, that poopie is something he should deal with, somehow.

DUMPSTER DIVING SCHISMS AND DISPUTES

Who "owns" dumpster diving? Who "invented" it?

In Hollywood, I had a gay associate producer give me a tongue lashing, informing me that gay people had pioneered dumpster diving and my book didn't give them any credit whatsoever! Dumpster diving, he informed me, is one of the primary sources of "art and style," and "oppressed" gay people pioneered the practice of "trashing parties" that raided the homes of Hollywood stars.

In an anarchist publication, I was taken to task and accused of all kinds of terrible things, but especially terrible was the fact that ANARCHISTS had pioneered dumpster diving and my book utterly ignored that.

A major scholarly book about garbage went out of its way to avoid mentioning my book, though it did mention the essay "On Dumpster Diving" by another author who mostly wrote about travels with his doggie and very little about dumpster diving except for that one section. Knowing very well that *TAandScDD* is the seminal text on the subject, it was conspicuous by its absence.

Undoubtedly, in my first book I should have mentioned the pioneering work of Urban Ore in California. Founded and managed by Daniel Knapp, Ph.D., Urban Ore combs the trash of Berkeley, California, for products and materials it can clean up or repair and then sell. Knapp made an interesting point in a 1995 essay which I would like to share:

Republicans don't own the philosophy of conservatism any more than Democrats own liberalism. Environmentalism, for example, has a deeply conservative streak derived from our view of nature. The natural world doesn't waste; it conserves. Discarded animal and vegetable matter never goes to waste. It is recovered by natural cleanup crews and processes and is converted into the next generation of resources.

Occasional catastrophes wipe out species. To avoid being a (further) natural catastrophe, we humans need to be conservative.

Also deserving credit is "garbologist" William Rathje, who wrote the book *Rubbish! The Archaeology of Garbage* with Cullen Murphy.

There should not be merely one dumpster-diving book in the world nor, for that matter, merely two. Let there be specialized books about, for example, gay and anarchist dumpster divers and their particular struggles and triumphs. Let stories about dumpster diving be told in every form of media, always with the intent of pulling our civilization upward . . . nothing wasted, nothing thrown into the pit of perdition, a civilization moving out of the trashy abyss toward the heavens.

But first we have to struggle and deal with conflicts and schisms, and if a handful of documents slices the fat, piggy throat of evil corporations, exposing them to a pack of media jackals, so be it.

Not to destroy . . . but to reform, and pull the civilization upward, ugly and smelly though the process may be at times. That's just part of life.

Dumpster diving is a kind of elevated consciousness that extends far beyond the actual diving of dumpsters, just as the outlook of a karate or judo master extends beyond actual sporting matches or street fights. For example, as an activist I would often file Freedom of Information Act requests for the e-mails of public officials, narrowing down my search by forcing them to turn over *indexes* of e-mails. I would literally prowl their electronic trash cans, looking for good memos. The process of gaining access and then sifting through data crap for nuggets was EXACTLY like dumpster diving, only quite ethereal, less physical and more mental. But it was the same process, and I was good at this process because of my "dumpster consciousness." Modern hunter-gatherer dumpster consciousness can be applied in all sorts of areas.

DUMPSTER DIVING A POLITICAL OFFICE

Arriving in Grand Forks, North Dakota, in a yellow Ryder truck, wearing Desert Storm camouflage pants and a black trench coat, pockets stuffed with the bric-a-brac of evacuation—beef jerky, maps, important phone numbers—I applied to law school and secured lodging in short order. I dumpster dived a black futon, kitchen table and chairs, and a few more filing cabinets . . . why drag furniture all over the country when people are always throwing away perfectly good stuff? Why buy a table when you can always dive one? Though my marriage had dissolved, I was still near my precious child, and a part of his life, and I had a plan for the future. I also had a 1986 Ford T-Bird that my brother Slash sold me for $50, God bless him.

I had uprooted myself with confidence, knowing the cold, hard universe would provide, as it always does, from dumpsters.

Soon I was back to my old activist tricks, joining a protest in city council chambers against a lopsided street tax assessment. One of the aldermen for my city ward, Jerry Lucke ("Lucky"), told the crowd of citizens they shouldn't be whining so much, that the average tax assessment was "chump change."

The keen senses of the diver know when something good is about to fall to the ground. I stared at Jerry Lucke hard after that utterance and said to myself, "There is a man who isn't going to run for re-election." I gathered signatures and declared myself his sole opponent in the race, making it clear I was a professional dumpster diver who had been arrested a few times in protests.

And after the deadline to get on the ballot had passed . . . Jerry Lucke declared himself out of the race! He had been hired outside the state and couldn't turn down the high-paying job. My instinct paid off. He didn't want his office and wouldn't stick around to defend it.

After living in Grand Forks, North Dakota, the bare minimum amount of time to run for political office, I assumed power and, well, actually became famous. My doings were newspaper headlines and television stories practically every other day, as a clique of furious conservative businessmen announced a recall effort. And I played my office for all it was worth, opposing a new landfill and promoting recycling. As a new bookstore opened in our refurbished and tax-subsidized downtown, *TAandSoDD* was placed prominently in the window. My suggestions to the Department of Engineering about reform measures aimed at curbing solid waste had the force of committee referrals. Wearing a little white hardhat and a sky blue necktie provided by my mother (originally dived out of a dumpster), I toured the landfill before and after little political soirees with gouda cheeses and Vienna sausages.

Ah! I thought. This is the life! This is, in some ways, even better than Seattle.

I had, in effect, "dumpster dived" a political office.

Which brings me to a new idea and tangent. I call it the Hoffman Model of Effective Political Sojourning. It occurred to me that, but for my little son, I would never have left a life of radical activism supplemented by dumpster diving in a major metropolitan area like Seattle, where life is good, wages are high, and there are Big Issues That Really Matter. After all, the city of Seattle is bigger than the entire state—I'M TALKING THE ENTIRE STATE—of North Dakota. There are BUILDINGS in Seattle that see more people in a day than all of Grand Forks. And yet I was forced to come here, to an environment I would never choose, where I found out that a single committed and skillful big-city activist can turn the town upon its ear and, yes, even gain a powerful political office.

How well I recall knocking myself out—KNOCKING MYSELF OUT AS A BIG-CITY ACTIVIST—to get a few lines of copy in the *Seattle Times* about this or that important issue. And how happy we would be to get those few lines instead of having our demonstration ignored by the media, as it so often was. But in little towns of 50,000 or so, where so little actually happens and people are afraid to embarrass themselves in front of neighbors they have known their whole lives, media manipulation is shockingly easy. You can stand in front of a school, for example, protesting military recruiting . . . and whoever heard of somebody doing THAT out in the boonies? Why, it's a front page headline! By simply going around to small towns in a van, sending press releases out before you like angelic heralds, you can "dumpster dive raised political consciousness" in micro-metropolitan areas.

It is just too easy! And who needs to pay for food along the way? That's what grocery store dumpsters are for, after all. But perhaps you don't need to travel like a nomad, taking off time from work, when you can obtain an affordable little summer cabin near your chosen micro-metropolitan target area for, like, a few hundred dollars.

DUMPSTER DIVING A LITTLE HOUSE ON THE PRAIRIE

There are thousands of abandoned farmsteads in North Dakota, thousands of abandoned commercial buildings in tiny towns—often outnumbering the *occupied* houses and commercial buildings—and, at last count, 400 abandoned country churches.

The place where I actually live at the moment I purchased for $3,500 with $890 up front and years to pay off the rest. It has a nice yard, and the previous family left a lot of stuff, including a Christmas tree decorated with candy. I have written in another context (a cartoon in the Loompanics Unlimited catalog) of the possibility of visionary homesteaders taking over dying little Midwestern towns for pennies on the dollar, from the buildings to the government to the laws, renaming the towns with subversive themes to reflect what is going on.

Lesbos, North Dakota.

Vegan, Montana.

Nude and Natural, Iowa.

In the case of Slope County, North Dakota, which has a population of approximately 950 and is the least populated county in the continental United States, well, this is a place that just *piteously cries out* to be taken over by weird, wonderful environmental elements who will, for example, paint the fire engine with polka dots and erect a statue of Julia Butterfly on Main Street. Also, North Dakota has the most liberal recall laws in the nation, so there's no need to wait for election time when you've got enough people to take over a political office like mayor or the entire city council.

In fact, I'll try to be of some help, OK? Here's the deal. Any environmentally oriented radical types who wish to go scouting for cheap land in North Dakota for their own weird, wonderful homesteads are hereby invited to use two of my buildings for as long as you require, and make any improvements you feel are needed, other than burning the buildings down or tearing down the walls.

Oh, and please don't plant any hemp, since that would cause problems, you know? If others come along bearing this invitation in their hot little hands, well, you are all going to have to share and work things out. I won't have time to mediate or visit you, but please drop me a line before, during, and after with a full report on your adventures. Be good guests or I will ask you to leave, but that doesn't mean you shouldn't share your *unique point of view* with the locals. By "good guests," I don't mean minor acts of rudeness which are inevitable in places like this; I mean, you know, don't rape, pillage, burglarize, commit arson, etc.

Also, this invitation doesn't extend to individuals who have been convicted of a sex crime. (As for other crimes, well, you're sorry, right? OK, good.) I reserve the right to reject anybody I deem unsound, unsuited, or unfit for the stated goals of this experiment.

The first building, pictured in the tall grass with this image of my son at the age of three, is in Inkster, North Dakota, a town that is filled to overflowing with abandoned buildings. As part of a widely publicized Dada art-inspired prank, I dubbed it the Dakota Devastation Art Museum and declared its fate symbolic of the fate of North Dakota. It's on the first dirt road behind the town's lone abandoned grain elevator (you really can't miss it), and if you find yourself out that-a-way, no, I do not own the abandoned gold-colored house right next to it. Neither of these buildings currently have electricity or running water.

The second building is in Alsen City, and it's near the highway. You can't possibly miss it either. If somebody has locked this building with a padlock or nailed it shut, well, that wasn't me, so feel free to break inside if you must. The stunt with the painted letters—another successful media event—was inspired by a similar sign that appeared in Seattle decades earlier.

Additionally, I own two vacant lots that I've never seen, which you are also welcome to use, if you can find them by obtaining a plat map with the county auditor and tracking them down from the coordinates, which are as follows:

E33′ of S76′ of E30′ of N80′ of S156′ & E33′ of S18.7 of N144′ Block 2 Valli Hai Addition of Minot City, Ward County.

North 20′ of Lot Fifteen, Block 8, Hannah Village, Cavalier County.

Be creative and consider what you would do with an old building or a vacant lot purchased for a pittance. Me and my little brother used to obtain cool stuff from the dumpster and look at it and think . . . Well, gee, we've got this really cool piece of junk. What sort of prank can we play with it? It's the same deal with old buildings and vacant lots. Give me something old and cool like that and, damn, I'll come up with half a dozen colorful stunts.

Alexander James Hoffman, age three, in front of the Dakota Devastation Art Museum, Inkster, North Dakota.

Say you've got a building that is so old and decrepit that it really will fall down any moment. What can you do with THAT? Well, first get a proper permit to burn the building in question (you may not even need one; just inform the fire department of your intentions). Once you've covered your bases as though you were going to do something *normal,* send out a press release that you are going to burn a building in order to symbolize driving out the evil of corporate agriculture from the Great Plains. Play loud weird music at sundown and dance, half naked, around the flickering flames, chanting incantations while your friends videotape it all.

What fun. What a stunt. When you're done you can announce that you own another five

Andrew Johnson, left, and Jon Towne, right, members of E6 Pranksters, in Alsen City, North Dakota.

buildings and you are going to burn them down, too. Who cares if it is true or not? Small town journalists will almost surely swallow your line. If you did something like this in the big city, you would probably be ignored. In little towns they will not only give you headlines, they will *editorialize*. They will not only editorialize, but you will become an enduring symbol, brought up time and time again in conversation and letters to the editor. *Fellow North Dakotans, we've got to get "value added" soybean processing in our country to turn the local economy around. We must do this soon, before it is taken over by characters like Stan Stankewitz who want to burn things down for sick supernatural kicks and for cat-screeching tracks on "punk rock" albums!*

OK, here's another one. Buy an old church and, in the same town, a building chock full of junk. Take the junk and, with a lot of sweat and artistic ingenuity, turn the church into a shrine to the Dumpster Goddess. Explain to local media why you feel compelled to offer this act of devotion. Urge others to come and make offerings of junk so the Goddess will smile upon them, too. Urge others to obtain roadside property and raise up shrines as well, throughout the Midwest, where devotees can not only make offerings but rest their weary bones while searching for homesteads. (Hey, I can't do this all by myself! I've kicked in two houses and publicized their locations. Do as the Master has done and, in fact, do *more* and *better*, like Mark the Big-Hearted Texan.)

OK, here's the last one. A vacant lot. Boy, a

vacant lot is tough. But one possibility is to raise up your own personal "monument." *Upon this very spot, May 1, 2005, Jim and Stacy Smith had incredible sex and then spent about an hour making fun of unenlightened local people who grow beer hops under contract for evil multinational corporations.*

I believe very much in the northwest corner of North Dakota—an area 50 to 75 miles north of my location, especially Cavalier County—as a future hotbed of "green libertarian" weirdness. Winnipeg is the most wonderful, cosmopolitan, BIG city, and it is only about a 100 miles north of that area. With old farm houses in the northwest corner of North Dakota selling for a pittance, with plots of land all over the place up for grabs and more going on sale every day, this area could become like Vermont at the end of the 1960s.

I think old churches are cool. Old country schools and wooden grain elevators are cool, too. For six months of the year, when it is temperate and not covered by snowdrifts and subarctic cold, North Dakota is kind of nice.

So, how did I obtain property for a song?

Dumpster consciousness, baby.

It is possible, even in big cities, to find cheap property by checking the rolls for "tax forfeited" homes. These are properties that, in most cases, have been empty and neglected for so long that nobody bothered to pay the taxes until finally the county in question just takes possession and tries to sell the home or land to whomever will pay the accumulated taxes. In major metropolitan areas, somebody with a lot more money than the average reader will snap up properties like that and try to turn them into rentals, or they might tear them down to "create" (I use the word loosely) a parking lot where fees can be charged.

In any case, even tax-forfeited property is out of reach, economically, for most young bohemians in the big city. And yet, incredibly, a "summer cabin" in a Midwestern state may be well within reach, a place to retreat for long vacations when the city is hot, or to hide out when some other form of "heat" becomes oppressive. In the Midwest, radicals might escape from the slavery of rent to complete their Master's thesis and find their center before returning to their metropolitan struggles. They might plant flowers and (as I have demonstrated) even assume political offices in a kind of private political playground that could change the very demographics of social struggle in this nation. And it all begins with the tax-forfeited rolls of property in little Midwestern counties . . . .

Back to the evils of the WTO, for those evils are relevant here. American agriculture is in a state of acute decay from which it will, in all likelihood, not recover due to the globalization of agriculture. The average age of an American farmer is 52, and young people are leaving rural areas just as quickly as they can earn college degrees and/or throw everything in the back of their vehicle and drive to Seattle. What this means is . . . empty properties on the tax forfeit rolls that can be snapped up for a song, especially in the plains states. Yes, you too can have a cabin where you can type out a manifesto! And it will cost you just pennies on the dollar!

What a country!

Most Americans spend about 40 percent of their income on housing. That's just too damned much. The need to pay for housing keeps us slaves. I grew up on a farm, and all my life paying rent has been hateful to me. I've managed to avoid it a number of ways. I've lived in a dorm and then gotten my student loans paid by the government for a tour of military duty. I've lived in housing in a national park. I've been an apartment manager and gotten my rent paid as part of the job. Now I live in a house that I bought for a song off the tax rolls.

Getting your own housing is very important. Imagine if you could somehow *take over housing or space for housing* in the big city, where the dumpsters are rich. This is not an easy task, but I say *this is what you must seek.* If you can only get away from the need to pay rent, the green steel horn of plenty will provide the rest. Even if you are quite some distance from the city, a drive to the city and concentrated diving can provide for your needs. It is possible to push up from a crack in the sidewalk and become a mighty oak, spreading your acorns far and wide.

Just Dive It.

THE FUTURE OF EARTH—GARBOSPHERE II

Earth being the original garbosphere, this project will therefore be called Garbosphere II.

Imagine a flat plain near a landfill, or perhaps a parking lot near a big city campus. Six brave volunteers . . . garbospherians, but we can just call them all "garbos" . . . will be left alone—AND STARK NAKED—upon the plain or the parking lot, where they will be expected to construct a world for themselves, using nothing but money, energy, and materials obtained from garbage.

So they could, for example, collect aluminum cans and buy something, like a word processor, upon which one of them can write out stories to be printed on dumpster-dived sheets of paper, which he can sell for money and buy, what? Maybe decent shingles for the buildings they've constructed from scrap lumber. In any case, the garbospherians will be extensively documented as they struggle to create a world from the waste of the world around them. What amazing prime time entertainment it might be, I should add. Maybe two groups can compete for a million dollar prize, their efforts to be judged by a building inspector to determine who has built better accommodations for themselves, thereby winning the prize.

DUMPSTER DIVING LOVE, IMMORTALITY, AND SALVATION

There is a story from an old Indian text that I am probably mangling all to pieces, but it goes something like this.

A young man was crazy for a beautiful woman. He pursued her night and day, trying to seduce her. Finally the young woman said, "In order to make you happy, I will give you my beauty. Only allow me a few weeks to prepare it for you."

In those few weeks, she forced herself to vomit, over and over again, and saved all the vomit in jars. Then, pale and gaunt, with dark circles under her eyes, looking like a horrible specter, she called the young man to her house and pointed at the vessels containing the vomit.

"Here is my beauty," she said.

I guess that I'm not going to Krishna Heaven, because I take this lesson and apply it THE VERY OPPOSITE OF HOW IT WAS INTENDED by the person who told me the story upon a sidewalk in Seattle, dressed in a saffron robe. The moral of the story is supposed to be "beauty is all shit, beauty is an illusion, when you fall in love with beauty, you fall in love with shit, therefore become a monk and devote yourself to Krishna, for this is higher and better."

But my take upon the ancient lesson is this: my word, THAT explains why I am so good at picking up girls, why I have known love so often, and enjoyed it so much! Because it's just like dumpster diving!

These things I thought to myself while waiting to score free Hare Krishna food at one of their feasts and, by the way, if you've ever eaten at a Krishna feast, you have eaten dumpster-dived food, praise be to Krishna Krishna Krishna.

At that moment a swami came around, giving each of us a little drink of water from the Ganges River. And, applying my dumpster-diving outlook, I thought to myself, well, I don't know if this is going to work or not, but what the heck? Who wants to come back as a human being and have to go through potty training again? I'd rather be a cosmic butterfly flitting between galaxies, or one of those cool turtles that comes up for air every hundred years and puts its head in a garland of flowers, Krishna Krishna Krishna.

And when I went home that evening and told my wife about drinking the water in which Indian villagers bathe, cremate their dead, and flush their toilets, Tina was FURIOUS with me!

"You broke free of the eternal cycle of death and rebirth?!" she shrieked. "And didn't even call home so I could come and do the same? This is typical of you. You're such a self-centered pig."

"No," I said. "Not a pig, dear. I don't have to worry about being a pig, no more no more no more!"

THE FINAL SECRETS OF THE MASTER

The lessons learned dumpster diving can be applied to finding love, to perceiving the exact moment when an object of your desire is filled with desire for you, and saying to yourself, "This is a girl who wants me to ask her out." And there is a good reason why dumpster divers seem to have big families and, even between all those lines of weirdness, seem oriented toward "family values," by which I mean . . . .

They tend to fall in love and reproduce, and to be very oriented toward love and reproduction, and they use dumpster diving to support what is most important, which is their family unit, no matter how unorthodox that family unit might be. It is clear by the way so many fans reach out in a beautiful, vulnerable way, that something very TRIBAL is happening, the emergence of an ultramodern hunter-gatherer culture in the very midst of urban civilization, a culture complete with rituals and magic, with legends and myths and its own gods and goddesses, gurus and students ascending, themselves, to guru status, a great and eternal circle of life.

We are a brave and noble tribe of visionaries, we dumpster divers, and it is our destiny to pull civilization upward, for our children to walk upon distant planets where they will erect glorious, prospering kingdoms.

I am so filled with sadness to tell you that my model of a perfect, nuclear, dumpster-diving family presented in the first book has come asunder. Let me assure you it had nothing at all to do with dumpster diving. Much of it had to do with growing apart over time, with one partner becoming more and more conventional while the other became increasingly less so.

It was not easy to come home and inform a little girl from Crookston, Minnesota, who was carrying my baby that, a few moments ago, darling, I was walking by a dark building on my normal route to work, late at night, when suddenly a little red light was dancing upon my chest, very scary shit, so I ran! But what was I going to do as a result? Crawl? Whimper? Back down in my quest to pry open the Police Intelligence Section of the Seattle Police Department? Not me, baby.

If you read the official after-action report about the Battle of Seattle, you will see the police blame their massive intelligence failure upon constant scrutiny placed upon their Intelligence Unit over the years, which effectively intimidated the unit from spying, effectively, on radicals. I am proud to have been a part of that long and personally costly effort. But there is a terrible price to be paid for the obsessive nature of an activist, somebody who would rather sift through hundreds of documents, looking for nuggets of dirt to drag through the pages of the alternative press, than make love or go out to dinner.

But the good news is . . . you can start all over again.

The tribe is eternal.

And, as every good dumpster diver knows, in the very heart of destruction, decay, and hopelessness are the seeds of life and planet-changing potential.

OTHER SEEKING, SEEKING THE GODDESS, SEEKING GOODIES

CULT AUTHOR EXTRAORDINAIRE SLASH LAW SCHOOL STUDENT SEEKS DARING, SINGLE BY CHOICE WOMEN WHO WANT TO HAVE BEAUTIFUL, SMART BABIES, NO STRINGS FOR ME AND NONE FOR YOU. I AM WILLING TO PROVIDE EXTENSIVE GENEALOGICAL DATA. YOUR BOUNDARIES HELD HOLY. OTHER WOMEN SEEKING NON PLATONIC FRIENDSHIP, EASY GOING ROMANCE COMBINED WITH THE SHARING OF IDEAS AND RESOURCES TO CHANGE THE PLANET FOR THE BETTER IN WEIRD AND WONDERFUL WAYS, PLEASE WRITE. ALL OTHERS, INCLUDING MY BEAUTIFUL BROTHERS AND SISTERS IN THE GAY AND ANARCHIST COMMUNITY, AND THOSE WHO WISH TO SEND IDEAS OR, BETTER YET, ACTUAL *RESOURCES* TO THE BUILDING OF WHATEVER THE HECK IT IS WE ARE BUILDING HERE, PLEASE WRITE TO:

John Hoffman
P.O. Box 169
Gilby, ND 58235

Gurus come and go, but the dumpster tribe is eternal.

ENDNOTES

1. See Rhyne v. Kmart Corporation., No 98CV5519 (Super.Ct., Gaston Co., N.C.).

2. Willie Fulgear, a professional scavenger, parlayed his recovery of 52 stolen Oscars into brief fame, a $50,000 reward, and a headline-grabbing invitation to the Academy Awards. Unfortunately, thieves stole most of the reward by making off with a safe from Fulgear's house. In another odd twist, Fulgear was the half-brother of one man who pled "no contest" to stealing the Oscars, which originally numbered 55. Anthony Hart maintained his innocence even after making his plea, saying that the system is not perfect and he didn't want to get caught up in it. While on probation and subject to all kinds of attention due to the alleged Great Oscar Heist, Hart got into legal trouble again when he was accused of "stealing" pallets, an item which frequently straddles the world between "stored on a loading dock" and "tossed aside because damaged but potentially still valuable."

Despite the connection with Hart, Willie Fulgear was never charged with anything. Fulgear's safe, as well as three of the original 55 Oscars, remain missing. All this information comes from *E! Online News*, "Oscar Heist, Case Closed" by Bridge Byrne, December 6, 2001, 4 PM Pacific Time.

Speaking as the world's other Hollywood dumpster diver, I would like very much to solve this mystery of the whereabouts of the three missing Oscars. These Oscars, wherever they might be, are too hot to be sold or even displayed to close friends. They will, in all likelihood, remain hot for hundreds of years and will carry with them the taint of thievery. If the person or persons in possession of these three little gilded hostages will mail them to me along with a statement of under 100 pages, 12 point type, addressing whatever you want to say to "The System" or "Hollywood," then I solemnly swear, upon receipt of all three statues, that I will make my best effort to have your statement published in such a way that it can be accessed and read by the whole world. I will make sure it is transmitted in its entirety, and I will comment upon it in a positive, open-minded, scholarly way, thus extending the credibility of my "dumpster cult author persona" over your thoughts and words.

I will be required, of course, to turn the Oscars over to the proper authorities as quickly as possible, along with your statement, but I will make a copy of your statement first.

Look at those little men, or think of their little faces if they are not there with you now. Don't they seem as though they would like to rejoin their friends and family, as much as they have come to admire you? A part of you will always be a part of them, will always possess them. In the very act of letting them go, you exercise your ultimate possession, power, and control over them. They will always be known as an expression of you, and they will carry your power everywhere they go. The safer they remain, the greater the expression of your power.

Hollywood is listening. The whole world is listening. And I, who have been in the dumpsters of Hollywood and stood in the limelight but have NEVER obtained such treasures as you possess, I am listening. Do not waste this moment. Send the little men on swift wings with some kind of communication to my mailing address listed at the end of Chapter 9. I will wait, patiently, for your longer statement and keep my promise.

3. As per the settlement agreement, I am NOT stating nor implying that the destruction of approximately 125 police internal affairs documents by the Seattle Police Department after failing to turn these documents over for nearly two years was a deliberate act. Rather, it apparently was a mistake.

Mistakes happen.

4. Text to this footnote was shredded by mistake.

5. Julia "Butterfly" Hill pulled off one of the most successful eco-protests in history, sitting in a California redwood for two years and six days on a plywood platform no larger than a queen-sized bed in order to save an ancient tree (dubbed "Luna") from Pacific Lumber MAXXAM. She ascended the tree on December 10, 1997, expecting to spend two to four weeks helping out as part of a two-person tree sit, with a safety line attached. Within a week, she undid her safety line and flitted about like a wood nymph. She chose the "forest name" of Butterfly and vowed that her feet "would not touch the ground" until she had done everything in her power to protect old-growth areas and "change consciousness."

Media attention followed, first in a trickle, then a roar. Butterfly was interviewed daily by cell phone on radio and television, sometimes spending six to eight hours a day on the phone. Television crews traveled to the tree from as far away as Germany, England, and Israel, and celebrity visits included musicians Joan Baez, Bonnie Raitt, Mickey Hart of the Grateful Dead, and the ironically named actor Woody Harrelson. After more than a hundred unreturned phone calls, Butterfly managed to establish a rapport with John Campbell, president of Pacific Lumber. Ultimately, they met and exchanged symbolic gifts.

A complex deal was eventually worked out which saved Luna as well as a buffer zone of trees around it. Victorious, the Goddess Julia Butterfly descended into the media firestorm she whipped up from cell phone calls and poems on pasta boxes. She went on to author two books, *The Legacy of Luna* and *One Makes the Difference: Inspiring Actions that Change the World.* These books are approximately 30 percent post-consumer waste and 70 percent Smart Wood Certified, and proceeds go to Julia Butterfly's Circle of Life Foundation, which reflects how small, attainable acts of individual responsibility and accountability can change the world.

Unlike activists who come off as dour and self-righteous, full of words instead of deeds, Butterfly's personality and two years of hard-core direct action in the face of sustained, powerful opposition won admiration even from her opponents. Today, she endorses actions rather than products, such as outdoor clothing giant Patagonia's move to use only organic cotton and recycled packaging materials.

Information for this endnote, including turns of phrase and whole sentences, was taken liberally from media articles gathered on www.ottermedia.com.

6. I have quoted, verbatim, obewan45@hotmail.com as quoted on the website <www.socool.com>.

7. In addition, a male teenager who was fishing in Whatcom Creek was killed. Initial and vivid media reports tended to emphasize the two 10-year-old victims through nothing more than dumb luck of the draw. I have lifted info and descriptive phrases about these events very liberally from the following newspaper reports: "Gas pipeline explodes," June 11, 1999, *Bellingham Herald* and "Bellingham fireball turned dreams of reviving salmon habitat to ashes," June 12, 1999, *Seattle Times.*

8. This sentence proves that Ludwig Wittgenstein is full of shittgenstein.

Wittgenstein, perhaps the most influential 20th century philosopher, put forth a theory that human language is the only thing in the world that exists—the only thing you could prove, at any rate. Wittgenstein built this theory upon an assertion that "private language" can't exist. Imagine, Wittgenstein said, if a man writes "S" in his diary whenever he has a certain sensation. This sensation has no natural expression, and "S" can't be defined in words. The only judge of whether "S" is used correctly is the inventor of "S." The only criterion of correctness is whether a sensation feels the same to him or her. There are no criteria for its *being* the same other than its *seeming* the same. So he writes "S" when he feels like it.

He might as well be doodling. The so-called "private language" is no language at all. The point of this is not to show that a private language is impossible but to show that certain things one might say about language are ultimately incoherent. If we really try to picture a world of private objects (sensations) and inner acts of meaning and so on, we see that what we picture is either regular public language or incomprehensible behavior. The man might as well quack as say or write "S." (Last paragraph quoted from <www.etm.edu>.)

And to all this I say *shittgenstein.*

In choosing his vague "S" word as an example to support his point, Wittgenstein draws too extreme an example. What about a new word invented by a writer? The writer invents the word and sets its definition. (Others may take the word and keep expanding or contracting its definition, but initially the inventor of the word does this.) In the moment between inventing the word and sharing it with another human mind . . . in the case of a word coined for this manuscript, a period of 48 hours between writing and e-mailing the revised manuscript to a publisher containing the word *ohre* . . . private language did indeed exist. The fact it was not an incomprehensible "duck quack" was proven when the publisher found the new word acceptable and coherent.

Wittgenstein also glosses over mentally ill patients who, while having incomprehensible elements, are not entirely bizarre or senseless. He also glosses over the very last living member of, say, a dying tribe who still speaks the language of that tribe, S.

And, as you might have guessed, *shittgenstein* is a particular form of bullshit which fills an individual (such as Wittgenstein) while he or she is asserting the "private language can't exist" thesis. Though, in the case of philosophy professor Donald Poochigian of the University of North Dakota, perhaps he is only asserting such a thesis in the course of teaching us what Wittgenstein was trying to express. Thus, *pseudo-shittgenstein.*

While I'm being all *philosophical,* I will express that some words I uttered in the video *The Ultimate Dive* were out of context and not an indication of philosophical weakness. I said that, while dumpster diving, one certainly finds a lot of cool crap to hang on walls, but what I liked were *functional* objects.

The fact that beauty and decoration is a *function* is something that smarmy self-defined and proclaimed artists have been using for quite some time to trip up anybody who utters these sentiments. However, these sentiments attempt to express quite a common feeling, that there is a *big difference* between the function of, say, an air conditioner and the function of a painting hanging upon a wall. It is artists themselves who have taken art and put it into museums and sanctified churches and special areas of walls where it is securely contained like a virus, and where people go, expecting that art resides in that place, and does not so much reside in other places, thus forcing the illogic of beauty and decoration as *not an ordinary function* upon the rest of us, and making ordinary people think of themselves as *not artists* and, might I add, *not writers, not philosophers, not spiritual leaders, not politicians, not nor not.* Thus art becomes an "anti-function," an inadvertent and uncontrolled destructive force, capable of alienating us and taking away our esteem when we come in contact with it, so that one might very well say one prefers "functional" objects over so-called "art."

These forms of inclusive and exclusive disjunction . . . indeed, all forms of inclusive and exclusive disjunction . . . trip us and trap us in language games. In that regard, that *language* and its pitfalls is what we struggle with all the divelong day, Wittgenstein was brilliant and had a point. The answer to this and many other philosophical dilemmas, which I explore in an as yet unpublished manuscript called *Loose Canons* (a.k.a. *Love Children of the Cartoon Cult*), is to embrace a system called "enlightened moral opportunism," which you will find a lot more fun than Wittgenstein's credo to "accept and endure." But, in the words of a basic assertion of *TAandSoDD*, "You are an artist, a creative being, 24 hours a day."

Everything you do can be art. Everything you do can pioneer style.

You are the leading edge of God's artistic urges.

Now go forth, find the materials you need, and multiply.

9. This had originally read "A capital building set against a flaming sky" in Hoffman's manuscript draft, submitted well before the events of September 11, 2001. Hoffman requested the sentence fragment be changed to the one you read. The original manuscript, with overwhelming proof of its pre-9/11 creation, remains in possession of the publisher.